Current Topics in Microbiology and Immunology

Ergebnisse der Mikrobiologie und Immunitätsforschung

Edited by

W. Arber, Genève · W. Braun, New Brunswick · F. Cramer, Göttingen · R. Haas, Freiburg · W. Henle, Philadelphia · P. H. Hofschneider, München · N. K. Jerne, Frankfurt · W. Kikuth, Düsseldorf · P. Koldovsky, Prague · H. Koprowski, Philadelphia · O. Maaløe, Berkeley · R. Rott, Gießen · H.-G. Schweiger, Wilhelmshaven · M. Sela, Rehovoth · L. Syruček, Prague · P. K. Vogt, Denver · E. Wecker, Würzburg

Volume 42

Springer-Verlag New York Inc. 1968

Insect Viruses

Edited by

Karl Maramorosch

With 34 Figures

Springer-Verlag New York Inc. 1968

*Karl Maramorosch, Ph. D., Boyce Thompson Institute for Plant Research,
1086 North Broadway, Yonkers, N.Y. 10701/USA*

ISBN-13: 978-3-642-46117-0 e-ISBN-13: 978-3-642-46115-6
DOI: 10.1007/978-3-642-46115-6

Softcover reprint of the hardcover 1st edition 1968
 Title No. 4690.

Introduction

This volume contains seven chapters, based on papers presented at a Symposium on Insect Viruses, held in conjunction with the 67th Annual Meeting of the American Society for Microbiology in New York, N.Y., on 30 April—4 May, 1967. The Symposium was organized to bring together outstanding workers interested in various areas of insect virology, and allow an opportunity for a discussion of problems, approaches, and methods that would lead to further progress in basic and applied research. One of the principal reasons for holding the Symposium at this time was the feeling that the divergent areas of research, up to now studied separately by entomologists, medical and public health workers, geneticists, and plant pathologists, would be brought together, crossing the artificial borders and finding new, exciting and inspiring vistas. Insect viruses provide a rare opportunity to get acquainted with the work and methods of investigators in such related and yet distant fields.

Following the symposium, a decision was made to publish the papers in a single volume, extending the contents to provide a complete and scholarly review of each subject. Since viruses affecting insects have received little attention until recent years, it was felt that a fully documented presentation of diverse areas of insect virology merited publication. The invited authors, all recognized authorities in their respective fields, prepared their contributions in such a way that each is a concise unit. Typical viruses that attack and kill insects are discussed in two chapters. The first is by GORDON R. STAIRS ("Inclusion-type Viruses"), and the second by C. VAGO ("Non-inclusion Viruses"). Arthropod-borne animal viruses that infect both vertebrates and blood-sucking arthropods constitute, by far, the largest group of known vertebrate viruses, with over 200 different species described by 1967. This chapter ("Arboviruses") has been prepared by ROY W. CHAMBERLAIN.

The unique congenital carbon dioxide sensitivity of fruit flies, caused by a virus now known to be very similar to that causing vesicular stomatitis, but until recently studied almost exclusively as a genetic trait, has been covered by ROBERT L. SEECOF ("Sigma Virus of Drosophila"). The group of plant-pathogenic viruses that infect plants, as well as insects, sometimes causing fatal insect diseases, has been reviewed by KARL MARAMOROSCH ("Plant-Pathogenic Viruses in Insects"), who acted as the convener of the 1967 Symposium and editor of this volume.

An account of the increasingly popular techniques and achievements in insect tissue culture has been given by J. L. VAUGHN ("Insect Viruses in

Tissue Culture"). "Viruses—Living Insecticides" is the title of the last chapter, written by C. M. IGNOFFO.

The subject of insect viruses is of considerable scientific interest and importance, and one that appeals to an audience recruited from students of entomology, human and veterinary medicine, virology, genetics, biochemistry, zoology, plant pathology, microbiology, and other branches of biology. Those who attended the Symposium and participated in the formal and informal discussions realized how the details or aspects of a particular group of insect viruses may have a bearing on other groups. It is hoped that the publication of the revised and enlarged papers will add to the success attained by the New York City Symposium, by reaching a much wider audience of biologists and public health officials.

The Editor would like to pay special tribute to Dr. WERNER HENLE, who not only helped in the organization of the Symposium and arrangements with the publisher, but also showed remarkable skill and persuasiveness in obtaining the manuscripts. The Editor also wishes to thank the American Society for Microbiology for help in sponsoring the Symposium, and all of the contributors for their excellent contributions.

January 16, 1968
Yonkers, New York

KARL MARAMOROSCH
Boyce Thompson Institute
for Plant Research

Table of Contents

Inclusion-Type Insect Viruses. By GORDON R. STAIRS. With 6 Figures . 1

Non-Inclusion Virus Diseases of Invertebrates. By C. VAGO 24

Arboviruses, the Arthropod-Borne Animal Viruses. By ROY W. CHAMBERLAIN 38

The Sigma Virus Infection of Drosophila Melanogaster. By ROBERT SEECOF. With 2 Figures 59

Plant Pathogenic Viruses in Insects. By KARL MARAMOROSCH. With 6 Figures 94

A Review of the Use of Insect Tissue Culture for the Study of Insect-Associated Viruses. By JAMES L. VAUGHN 108

Viruses—Living Insecticides. By CARLO M. IGNOFFO. With 20 Figures 129

Subject Index 168

Inclusion-Type Insect Viruses

G. R. Stairs*

With 6 Figures

Introduction

The best known insect viruses cause the formation of inclusions in the cells they infect. They were first observed in the mid-nineteenth century and since then have received considerable attention from entomologists and microbiologists. In recent years, several authors have reviewed studies on this group of viruses (Bergold, 1958; Aizawa, 1963; Smith, 1963; Huger, 1963) and the present review will be concerned more specifically with structure and chemistry, the infection process, quantitative virus-host relationships and the natural occurrence of viruses in their host populations. The nuclear-polyhedroses of Lepidoptera and Hymenoptera (Sawflies), the cytoplasmic polyhedroses of Lepidoptera and the granuloses of Lepidoptera are discussed in detail.

Structure and Chemistry

Three major types of insect viruses cause the formation of inclusion bodies. These bodies are proteins and are not themselves infectious but infectious virions are usually occluded within them. One type of virus causes the formation of numerous many-sided (polyhedral) inclusions in the nuclei of infected cells. These inclusions are crystalline, very refractive, and are easily seen in tissue smears of squashed preparations examined under the light microscope. They are usually 0.5 μ to 15 μ in diameter. A characteristic name has been applied to this type and they have become known as *nuclear-polyhedrosis viruses*. Another type causes the formation of very numerous, minute granules in the nuclei of infected cells. These inclusions are also crystalline protein but much smaller than polyhedra and not so easily seen under the light microscope. They range in size from 0.2 μ to 0.5 μ and appear as small granules, hence, the name *granulosis viruses* has been applied to them. Both of these types are characterized by the formation of rod-shaped virions that contain DNA nucleoprotein surrounded by two membranes (Figs. 1, 2, and 3). The third inclusion-type virus appears to be quite different from these. Inclusion bodies are formed only in the cytoplasm of susceptible cells and the virions are spherical or subspherical

* Department of Zoology and Entomology. The Ohio State University, Columbus, Ohio

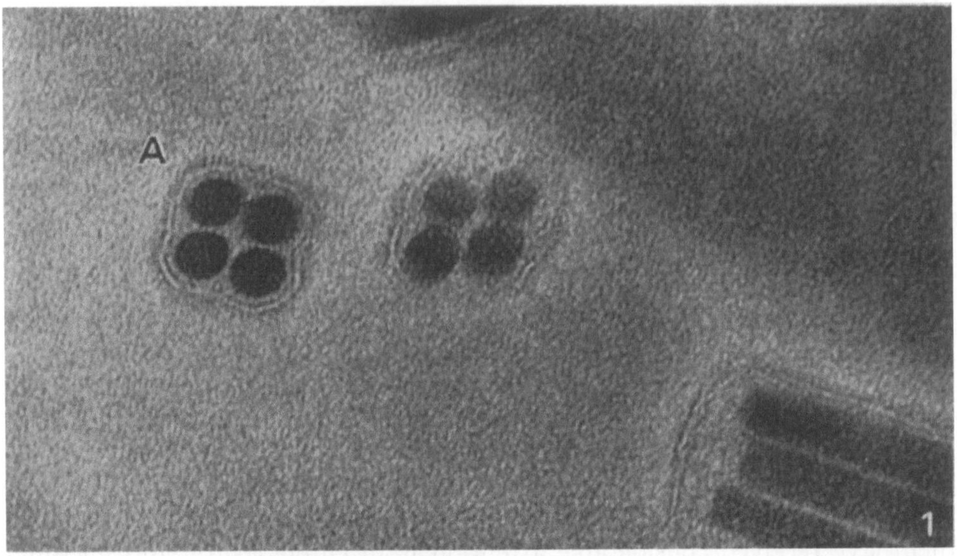

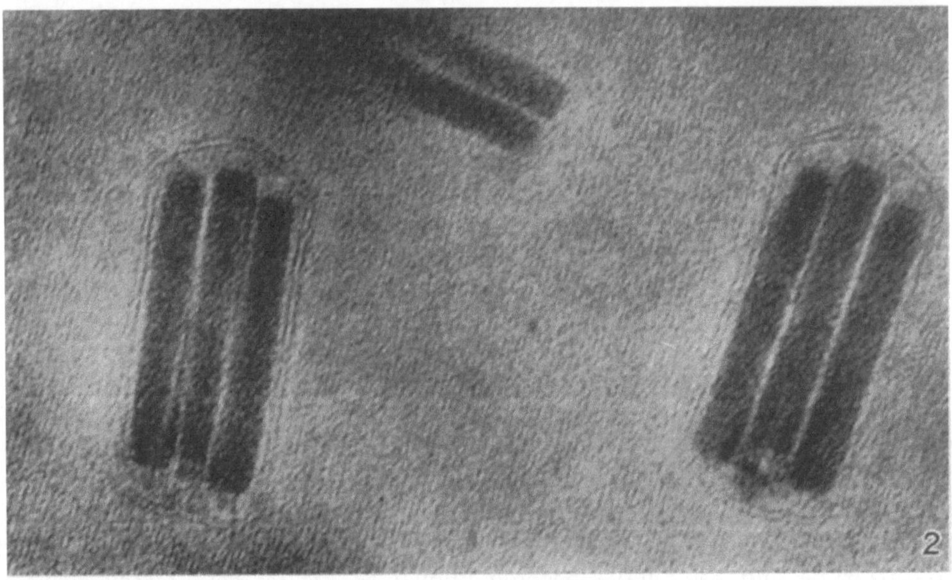

Fig. 1. A section through a polyhedron of the nuclear-polyhedrosis virus of the greater wax moth, *Galleria mellonella* (L.), showing (A) a cross-section of a bundle of virions and the associated membranes. 180,000 ×. Electron micrograph by W. B. PARRISH

Fig. 2. Similar to Fig. 1 but showing longitudinal sections through virion bundles. 150,000 ×. Electron micrograph by W. B. PARRISH

containing RNA nucleoprotein and no membrane structures. The inclusions, in which many virions are usually occluded, are crystals of protein with numerous flat sides. These inclusions are also called polyhedra and the viruses are known as *cytoplasmic-polyhedrosis viruses* (Fig. 4).

The inclusion bodies of all types appear to be of similar fine-structure and chemistry. In thin sections observed under the electron microscope the protein

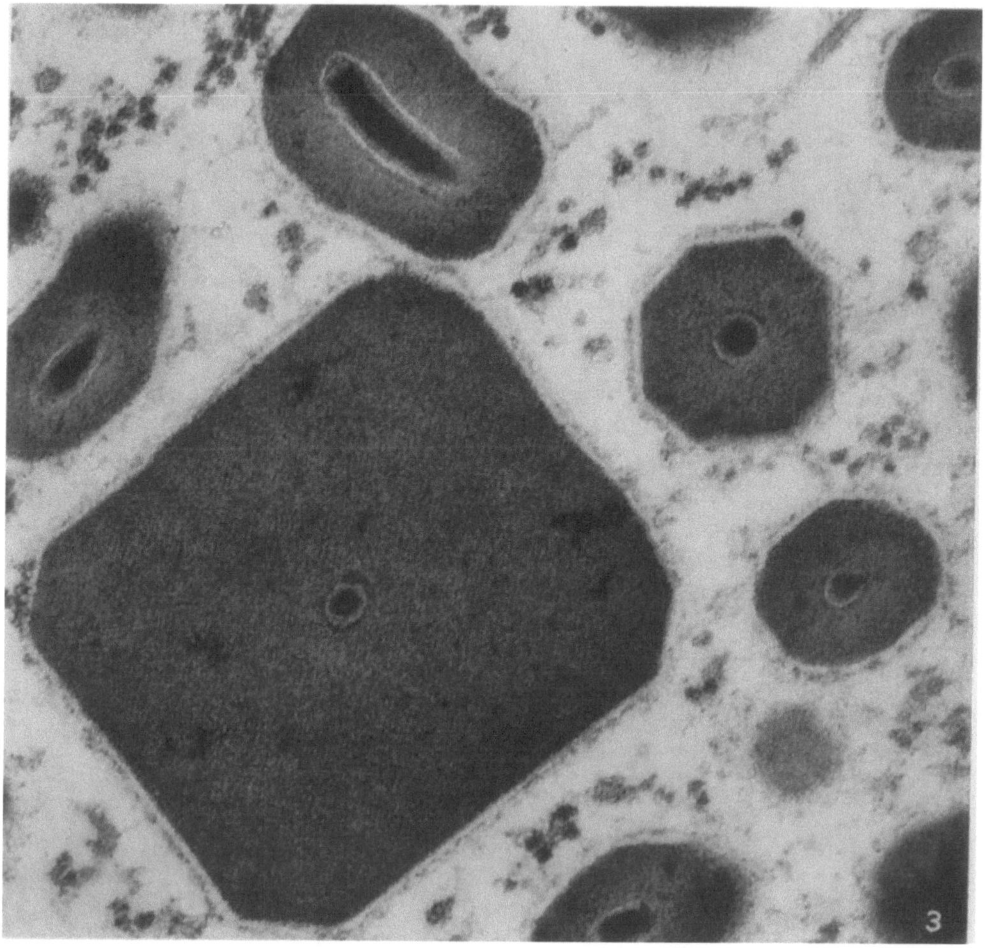

Fig. 3. A section of a cell of the codling moth, *Carpocapsa pommonella* (L.) infected with a granulosis virus showing individual virions in each granule. 100,000 ×. Electron micrograph by W. B. Parrish

subunits are arranged in very regular patterns (Bergold, 1963). Although the granulosis virus granules are not as large, their fine-structure is very similar to that of the polyhedra. The mechanisms of formation and crystallization of these proteins are not known but the sequence of events in relation to the development of the virions has been investigated. Apparently, the formation of mature virions and crystallization of the protein occur simultaneously in adjacent parts of the cell (Bird, 1964, 1966). An inclusion body continues to grow and occlude virions until it reaches a size of 0.5 μ to 15 μ. Each inclusion-body or polyhedron may contain many virions. On the other hand, a single inclusion is formed around each virion in the granulosis-type viruses.

There is some controversy concerning the control of inclusion-body formation within infected cells and recent evidence suggests the virion controls events leading to the formation of specific inclusions (Stairs, 1964). In the

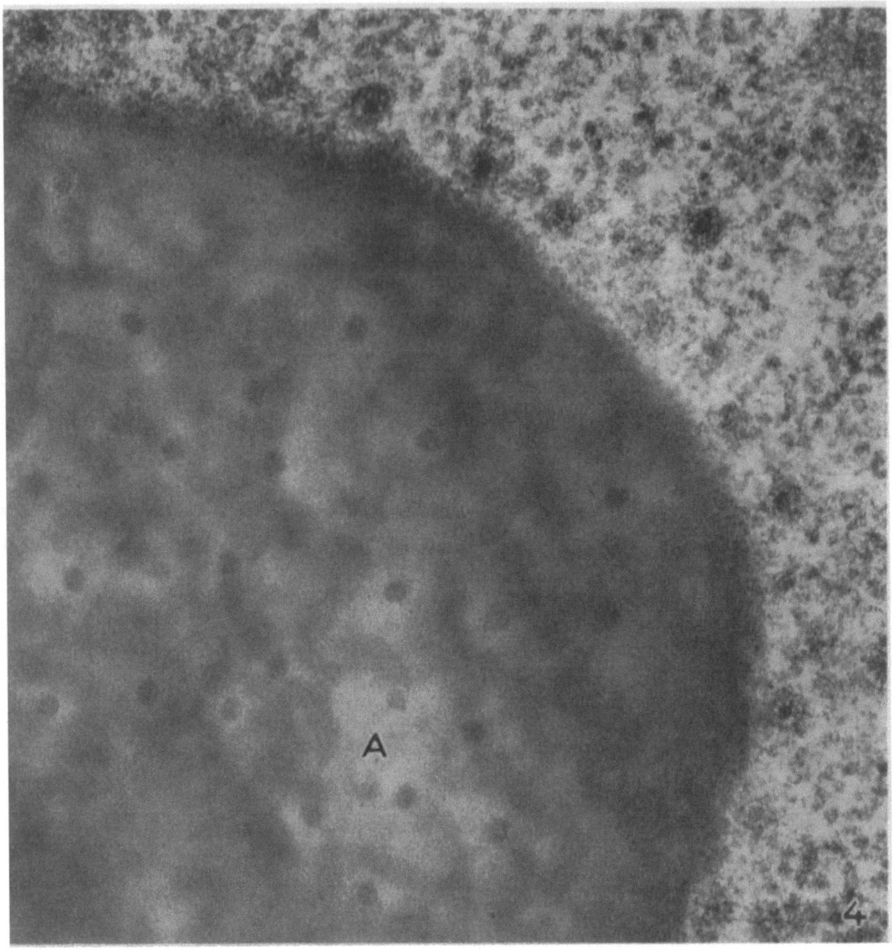

Fig. 4. A section through a polyhedron of cytoplasmic-polyhedrosis virus of the salt marsh caterpillar, *Estigmene acrea* (D.) showing the virions (A.) 50,000 ×. Electron micrograph by W. B. Parrish

granulosis viruses, many virions develop in a single cell-nucleus and the size and shape of each inclusion body may vary considerably. In some instances, the formation of large cuboidal inclusions and the more regular, small inclusions occur in the same cell (Fig. 3). A suspension of large cuboidal inclusions fed to susceptible larvae causes the development of cuboidal-type inclusions only. This strain was passaged through five host generations without reversion to the normal type. There was considerable variation in size but the shape was always cuboidal (Fig. 5). These cuboidal variants are also found in the granulosis viruses of *Choristoneura fumiferana* (Clemens), *Carpocapsa pomonella* (L.), and *Argyrotaenia velutinana* (Wlkr.) and further investigation may show that they occur commonly in many species of the families Tortricidae and Olethreutidae.

The first chemical studies of nuclear-polyhedra were made at the beginning of this century when it was observed that they dissolved in weak alkali solutions

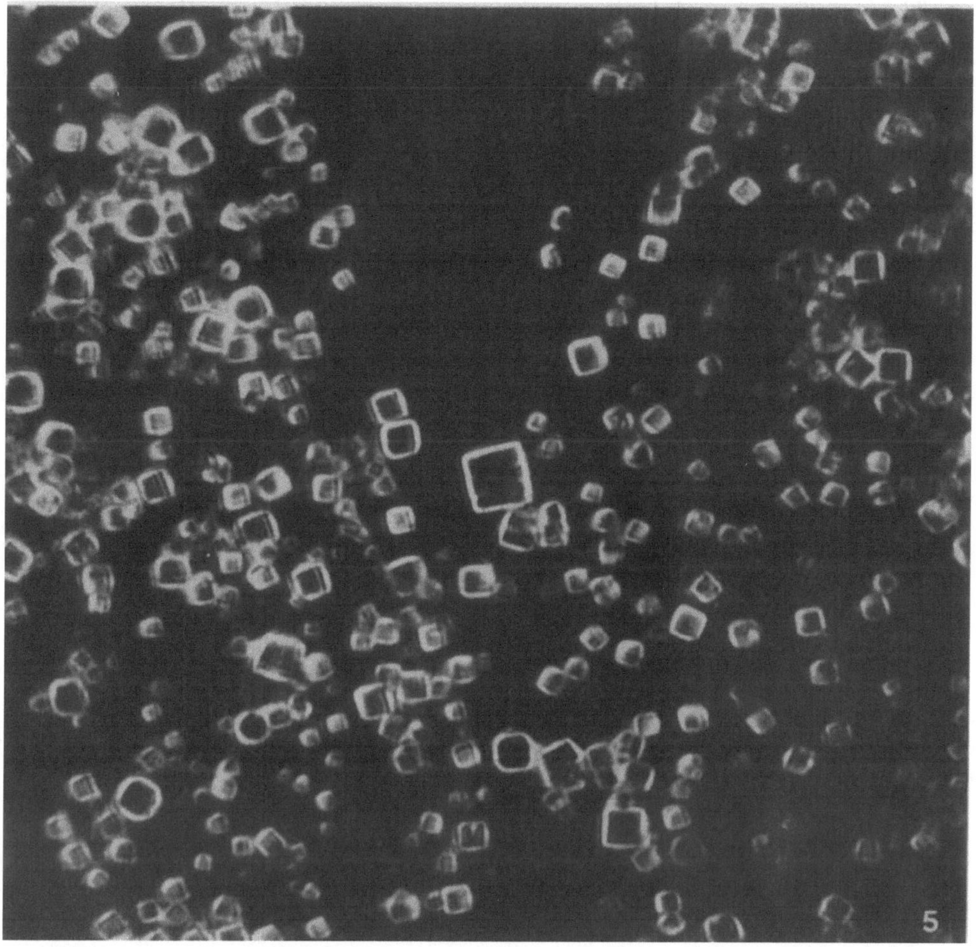

Fig. 5. A tissue smear from a larva of the spruce budworm, *Choristoneura fumiferana* (CLEMENS), infected with the cuboidal variant of granulosis virus. 2,200 ×. Dark field optics

but remained insoluble in water and were not decomposed by ordinary putrefaction processes (BOLLE, 1898). Much later, the infectious virions were released and isolated from polyhedra using essentially the same procedure (BERGOLD, 1943). Extensive chemical analyzes have shown that the inclusions consist of at least 19 different amino acids (WELLINGTON, 1954). The molecular weight of these proteins varies from species to species but ranges from 250,000 to 400,000 (BERGOLD, 1963). All inclusion body proteins have a very similar qualitative amino acid composition but the total quantity of amino acids varies slightly from species to species (BERGOLD, 1958).

In granulosis viruses the protein is very similar to that of the nuclear polyhedra (BERGOLD, 1958). The molecular weight of the granulosis virus of *Choristoneura murinana* (L.) is 300,000 and the total amino acid residue is 92.7 grams per 100 grams of protein.

The cytoplasmic polyhedra appear to be very similar to the nuclear inclusion bodies (Bergold and Suter, 1959). They contain a full complement of amino acids and exhibit a similar behavior in weak alkaline solutions but often require pretreatment with 0.1 % sodium hydroxide solution before they become dissolved in 0.1 % sodium carbonate (Kawase, 1964).

Recent studies have shown that both types of polyhedra contain considerable quantities of silicon (Estes and Faust, 1966). The function of the silicon is not known but it is believed to be important and may be responsible for the peculiar qualities of the protein crystals. Apparently, a combination of reducing agents and proteolytic enzymes function cooperatively in the host midgut to cause the dissolution of polyhedra and subsequent release of infectious virions (Faust and Adams, 1966).

The virions of the nuclear-polyhedrosis and granulosis viruses are relatively similar while those of the cytoplasmic polyhedroses are decidedly different. The former are rod-shaped, with dimensions of 20—50 mμ × 200—400 mμ and they are usually surrounded by two adjacent membranes while the latter are spherical or subspherical, without membranes, and range in size from 50—90 mμ (Hosaka and Aizawa, 1964). They contain RNA and no DNA, whereas the rod-shaped virions contain DNA only. Detailed chemical analyzes of the rod-shaped virions show that the core, as well as the membranes, have a fairly complete complement of amino acids (Wellington, 1954). In addition, the membranes contain ether-soluble lipids and the cores contain about 8% DNA. Analysis of the base composition of DNA from a number of nuclear-polyhedrosis and granulosis viruses show distinct specific differences between virions from the two groups (Wyatt, 1952; Bergold, 1963). This indicates that despite their morphological similarity, large differences in their nucleic acids occur.

Nuclear polyhedra contain both DNA and RNA (Faulkner, 1962) but cytoplasmic polyhedra contain RNA only and this RNA is probably all located in the virion (Aizawa, and Iida, 1963). The virions in the nuclear polyhedra, however, are believed to contain only DNA (Wyatt, 1952), thus, the location of the RNA in this group is not known. Thin sections of polyhedra examined under the electron microscope do not reveal specialized structures other than the virions, hence, the RNA may actually be a part of the virion. Possibly, it is contained in the two membranes that surround each virion

The nuclear-polyhedrosis and granulosis virus are very good antigens and extensive serological studies have revealed important relationships between inclusion-body proteins and virions from different host species (Krywienczyk and Bergold, 1960). The virions from one host species are generally not related serologically to those from another species unless the hosts belong to the same genus. In this case, the virions often cross-react very strongly. The virions from granulosis viruses react only slightly with those from nuclear-polyhedrosis viruses. In fact, viruses from each group causing disease in the same host are not related serologically. Further tests have shown that virions are unrelated to the inclusion-body protein that surrounds them. This demonstrates that the viral nucleoprotein is indeed different from the inclusion protein but it does not preclude the possibility that the production of the inclusion protein is controlled by the virion and is, in fact, an intrinsic part of the virus itself. It is quite possible that the nucleic acid of the virion is capable of directing the formation of a number of proteins that are different from one another. Recent evidence definitely suggests that production of the inclusion body is controlled by the nucleic acid contained in the virion (STAIRS, 1964).

Although the virions from different species do not appear to be related to one another serologically, the inclusion-body proteins show a high degree of serological similarity. This is particularly evident among the nuclear polyhedra of the Lepidoptera where all eleven proteins tested showed very distinct cross-reactivity with one another (KRYWIENCZYK and BERGOLD, 1960). Such a pronounced similarity probably does not exist between the proteins of the host insects. This suggests that the nuclear-polyhedrosis viruses of Lepidoptera are a closely-related group of obligate parasites with a flexible evolutionary potential. In other words, the similarity of the proteins indicates a common origin and a survivial course that has depended, and continues to depend, largely upon the presence of these proteins. The dissimilar serological reactions of the virions, however, indicate that for successful infection of cells of different hosts a high degree of protein specificity has evolved. If the virion-inclusion complex is viewed as an organism whose function is the propagation of itself all the proteins are probably important but those involved in the actual penetration of cells may be more essential to survival and, therefore, they have evolved more rapidly and specifically.

The nuclear-polyhedrosis viruses from the Hymenoptera (SAWFLIES) do not show any serological similarities of virions with those of the Lepidoptera. There is, however, slight similarity between their inclusion proteins suggesting that they are distantly related. The granulosis viruses of Lepidoptera show no reaction with the nuclear-polyhedrosis viruses of Hymenoptera but their inclusion proteins appear to be distantly related to those from nuclear-polyhedrosis viruses of the Lepidoptera. Apparently, these DNA, rod-shaped, viruses of insects are actually related and have a common origin. Serological evidence suggests three distinct lines of evolution, 1) the nuclear-polyhedrosis viruses of Lepidoptera, 2) the nuclear-polyhedrosis viruses of Hymenoptera, and 3) the granulosis viruses of Lepidoptera.

All of the preceding studies were made using purified preparations of inclusion bodies and subjecting them to weak alkali treatment (0.005 M to 0.03 M Na_2CO_3 + 0.05 M NaCl) to dissolve the protein, remove the membranes, and release the virion cores. Such analyses undoubtedly do not reveal the whole or true story. For example, sections of infected host tissue examined under the electron microscope show that many virions do not become occluded in the protein crystals and some do not even become enclosed in a membrane. The structure, composition, and function of these free virions should be thoroughly investigated.

The Process of Infection

Host insects may become infected in a number of different ways. Virus is commonly transmitted from adults to progeny. The infected progeny die contaminating the food plant. Healthy hosts ingest this food and become infected. Often hymenopterous parasites sting infected hosts, then move to healthy hosts. In this way, virus is injected directly into the host's hemocoele. Each of these infection routes is slightly different from the other but the virus develops in a characteristic manner regardless of how the host acquires it.

The rod-shaped, DNA viruses of Lepidoptera develop primarily in the nuclei of fat-body cells, tracheal epithelium, epidermis, and blood and only occassionally in the midgut; whereas, the spherical ,RNA viruses have been observed in the cytoplasm of midgut cells only. The DNA, rodshaped viruses of Hymenoptera also develop exclusively in the midgut cells but in the nuclei instead of the cytoplasm.

When a lepidopterous larva ingests nuclear-polyhedrosis virus the polyhedra usually remain intact until they enter the midgut. In the silkworm, *Bombyx mori* L., they disappear after being in the midgut for only 20 minutes and moving only one-tenth the length of the gut. The course of infection has not been traced beyond this stage but virus appears in the susceptible tissues four or five days later (STAIRS, 1963). How the virus reaches the hemocoele and thereby gains access to susceptible cells is not known. It is known, however, that a good deal of the ingested virus is inactivated in the midgut, and apparently, none is excreted (AIZAWA, 1962; STAIRS, 1963).

Lepidopterous larvae may also be infected by the injection of hemolymph of diseased larvae directly into the hemocoele or by injecting virions released from polyhedra by weak alkali treatment (BERGOLD, 1943). Generally, less virus is required to cause infection by this route than by ingestion (BERGOLD, 1947; STAIRS, 1965). When virus is introduced directly into the hemocoele and subsequently bioassayed, a typical virus multiplication pattern is observed. In the case of the nuclear-polyhedrosis virus of *B. mori* at 25° C, immediately following the injection of virions there is a decrease phase of about 10 hours followed by a logarithmic increase for about 40 hours (AIZAWA, 1959). The virus titer reaches 10^4 to 10^6 times the original inoculum and remains at this level until the larva dies.

During the first part of this process the cells do not show any cytopathic changes but at 30 hours infected nuclei begin to enlarge. The chromatin clumps in the center of the nucleus and shortly bundles of virus rods appear in the vicinity of the nuclear membrane (BOLLE, 1898; BERGOLD, 1943). At 48 hours small polyhedra appear and grow rapidly until the nucleus bursts open releasing its viral contents into the hemocoele. More and more cells lyse, the larva dies, then becomes very fragile and finally ruptures releasing the virus into the environment. A similar pattern develops following the ingestion of virus except that the period from infection to death may be slightly longer. Although the nuclear-polyhedrosis virus of *B. mori* has been used as an example here, the process is essentially the same wherever this type of virus is found infecting Lepidoptera (GLASER, 1915; STEINHAUS, 1949).

A similar pattern occurs in the Hymenoptera where the nuclear-polyhedrosis viruses attack the midgut cells only (BIRD and WHALEN, 1954). In this case, the virions released from ingested polyhedra have relatively easy access to the susceptible tissue. The first cytopathic changes of the infected cells occur in about 30 hours. The nucleus enlarges and the chromatin clumps at the center of the cell. Then polyhedra fill the nucleus and the cell finally lyses releasing its contents into the gut lumen. When the insect dies it becomes fragile, easily ruptured, and its viral contents are readily spread over the food plant (BIRD, 1961).

The infection process of the granulosis viruses is similar in many respects to that of the nuclear-polyhedrosis viruses. The infected nucleus enlarges but the chromatin does not become concentrated at the center. Instead it (or similar material) appears to form long strands that extend throughout the nucleus (HUGER, 1960). These strands are definitely Feulgen positive and they appear to be intimately associated with the development of the virions. They often persist in the nucleus and sometimes in the cytoplasm after viral development appears to be completed. In many instances, infected cells do not appear to lyse as quickly as do those infected with nuclear-polyhedrosis virus, but in other cases, lysis occurs very rapidly and larvae are easily ruptured. This difference may be related to the tissue sites of infection. Often these viruses attack only the fat body (HUGER, 1963), hence, an infected host may remain alive for an extended period of time but if the fat body and epidermis are both infected the host will probably be killed much more quickly. This appears to be the case with the granulosis viruses of *Choristoneura fumiferana* (CLEMENS), *Pieris rapae* (L.), *Pieris brassicae* (L.), and *Carpocapsa pomonella* (L.) where both tissues are susceptible to virus attack.

The granulosis viruses have not been studied as intensively as the nuclear-polyhedrosis viruses. As a result, much less is known about the infection process. They are not easily observed using the light microscope and their small size has probably discouraged investigation. Studies that have been made, however, indicate this group of viruses will yield valuable information regarding insect viruses in general.

A good deal less is known about the infection process of the cytoplasmic-polyhedrosis viruses. The midgut epithelium appears to be the only susceptible tissue and the inclusion bodies and virions are formed in the cytoplasm of these cells (SMITH, 1963). Upon infection the cytoplasm becomes filled with inclusion bodies and, as more cells become infected, the entire midgut becomes chalky white. The host may cease to feed and become shortened. Hosts usually remain intact after death, thus preventing the virus from being spread mechanically over the food plant. Infected larvae, however, often excrete virus with their frass thereby contaminating the environment before they die. This group of viruses is being studied intensively at the present time and much valuable information will be forthcoming in the near future.

During the development of both nuclear and cytoplasmic polyhedroses the host larvae undergo certain drastic physiological changes. The blood plasma becomes depleted during nuclear-polyhedrosis virus infection in larvae of *Peridroma saucia* (HÜBNER) (MARTIGNONI and MILSTEAD, 1964). In *B. mori* the total fluids appear to increase as well as phospholipids and fatty acids (KOMANO et al., 1966). Similar changes occur during infection of *B. mori* with cytoplasmic-polyhedrosis virus (KAWASE and YOSHIYUKI, 1965) where there is a general decrease of amino acids in the hemolymph and an increase of RNA and DNA in the midgut cells (KAWASE, 1965). The polyhedral protein in the nuclear disease of silkworm appears to be formed in the cytoplasm of infected cells, becoming crystallized only in the nuclei (KRYWIENCZYK, 1963).

Apparently, the formation of polyhedra requires a considerable amount of amino acids and the most accessible pool is the blood plasma. The specific virus proteins are formed in the cytoplasm and crystallize there in the case of the cytoplasmic polyhedroses. The same process probably occurs in the sawflies infected with nuclear-polyhedrosis viruses and in the granulosis diseases of Lepidoptera.

The inclusion-type viruses usually cause disease of the host larvae and they are also capable of causing disease of the pupal and adult stages (PAILLOT, 1926; BIRD, 1953; STAIRS, 1965). Larvae that die from nuclear-polyhedrosis or granulosis virus infections become pale, flaccid, and very fragile, so that they often burst open releasing the viral contents into the environment. If an individual host contracts virus late in the larval stage it will often pupate before a lethal infection develops. The pupal integument is usually much more rigid than the larval integument so the infected pupa does not become excessively flaccid or discoloured, hence the disease may go undetected in this stage. Similarly, an infected host may reach the adult stage before succumbing to the virus and, since many adult Lepidoptera and sawflies do not live for more than a few days, disease is often not detected in this stage. It should be emphasized, however, that the incidence of disease in the pupal and adult stages may be much lower than that occuring among larvae, therefore, detection is more difficult because a larger sample must be examined. In many cases, the infected adults mate and lay a full compliment of eggs. These eggs are usually viable and give rise to apparently normal larvae. Some of these larvae, however, develop typical virus disease and die during the early instars. It is in this manner that virus is most commonly transmitted from one generation to the next (BIRD, 1961).

The cytoplasmic-polyhedrosis viruses also infect pupae and adults. Heavy infections often cause adults to be malformed on emergence from the pupa case and the virus appears to interfere with oocyte development greatly reducing the reproductive potential of infected adults. Diseased adults are often much smaller than healthy adults and develop large, necrotic, tumor-like masses that are packed with the virus inclusion bodies (NEILSON, 1965). These viruses, however, do not always cause adult malformation or mortality. They appear to be transmitted to the progeny and there is some evidence of maternal cytoplasmic transmission (ARUGA and NAGASHIMA, 1962).

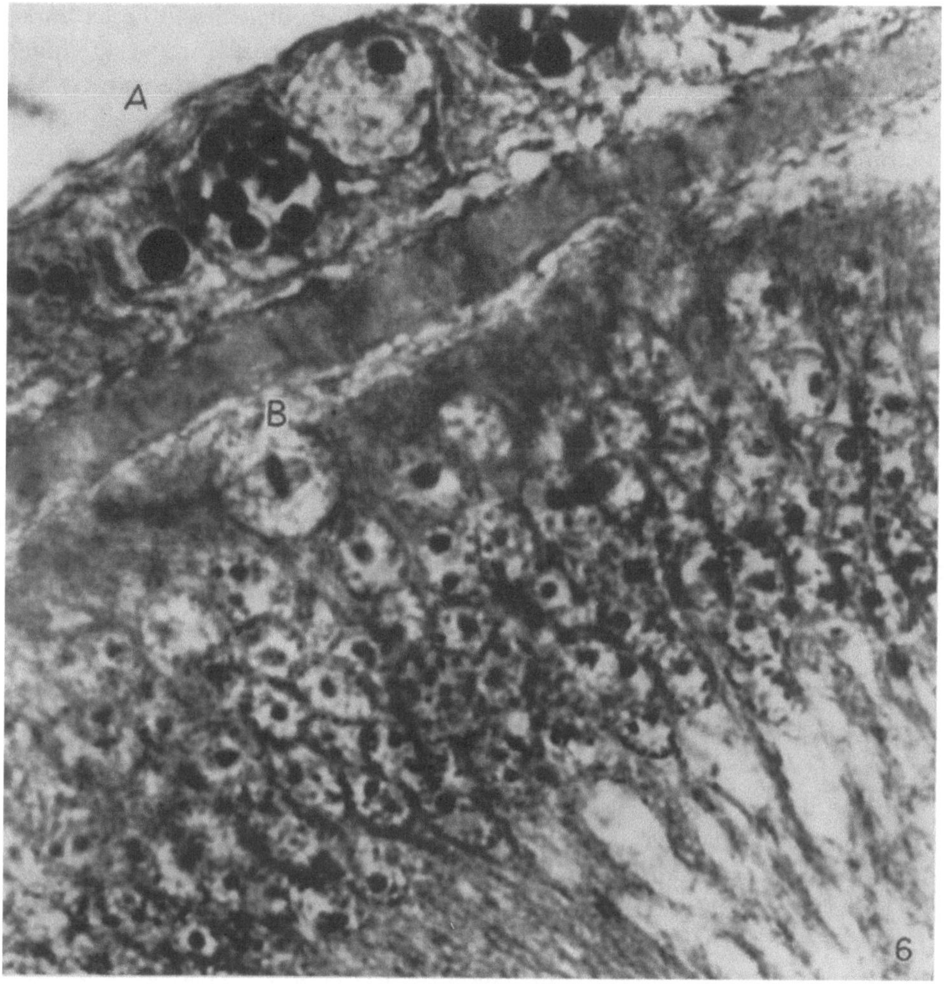

Fig. 6. A section through a wing-bud of a silkworm larva, *Bombyx mori* (L.), heavily infected with nuclear-polyhedrosis virus. Note polyhedra (A) in the cells of the peri-podium and the healthy embryonic cells in the wingbud which are still undergoing mitosis (B). 1,500 ×

One of the most interesting aspects of insect virology is the resistance of the embryonic regenerative cells to attack by the nuclear-polyhedrosis viruses (BIRD, 1953). In sawflies (Hymenoptera), the digestive cells of the midgut epithelium may be heavily infected but the regenerative nidi remain unaffected. When larvae begin to pupate a new midgut is proliferated and these cells remain non-susceptible until they begin to function as digestive cells. Often the host enters a prolonged diapause during this time that may last for a year or more. The virus, however, remains infective and begins to develop in the cells as soon as they become functional.

A similar situation develops in the Lepidoptera. If a larva contracts virus just before pupation it will undergo metamorphosis and die in the pupal stage but the incubation period (the time from infection to death) will be extended (STAIRS, 1965). A great many of the functional larval cells are very rapidly replaced by embryonic regenerative cells which are not infected and which become infected only after they differentiate (Fig. 6). Essentially, this is the beginning of a new infection so the insect survives for an extended period.

The mechanism responsible for the non-susceptibility of the embryonic cells is not known. Often, in larvae that are very heavily infected with nuclear-polyhedrosis virus, the only cells of the epidermis (or in sawflies the midgut epithelium) that are unaffected are those in the regenerative nidi. Since these embryonic cells are essential to the survival of the insect and, since the survival of the insect may be essential to the survival of the virus, the resistance of the embryonic cells probably has arisen to insure the survival of both the host and the virus.

This principle may also apply to the variation observed in the infection process from one virus-host association to another. The nuclear-polyhedrosis viruses cause infection in the fat body, blood cells, tracheal epithelium, and the epidermis of Lepidoptera and the midgut epithelium of sawflies. These infections appear to be very similar and the appearance of larvae at death is strikingly similar in that they become liquified. This suggests that host liquifaction and release of infectious viral material is of basic importance to the continued survival of the viruses. This also appears to be true for some of the granulosis viruses but not all of them. Certain granulosis viruses infect the fat body but do not infect the epidermis, with the result that hosts do not become fragile and liquified at death. In these cases, it seems that liquifaction is not so important to the survival of the virus and suggests these viruses have evolved other means of dispersion and survival in their host populations.

There is a good deal of literature indicating that certain inclusion-type viruses are capable of surviving in their hosts for an indefinite period of time without causing pathologies or cytopathic changes in the host cells. Most of these conclusions, however, appear to be based on observations that, at most, represent the considered opinions of the observers. A few workers have used the experimental approach and this approach has yielded different results depending upon the virus-host system studied. In most studies, the virus is assumed to be present within cells in an occult state, becoming activated or provoked to the pathogenic state by changes in the host cells caused by host adjustment to change or changes of the external environment (STEINHAUS, 1958; ARUGA, 1963; JAQUES, 1961; YAMAFUGI and HASHINAGA, 1966). It is also possible to explain most of these results by assuming a simple quantitative virus-host relationship. Assuming the external environment does cause changes in the host metabolism and the food being fed to test larvae is contaminated with a very low concentration of virus, it is possible that under changed conditions the lumen of the alimentary tract becomes more favourable to the survival of virus, hence, a higher proportion of hosts may become infected by the low dosage of virus. The quantitative approach has not been adequately applied to this problem but it is sure to yield valuable information.

To apply this approach properly, one must first assume that a virus-free host population can be selected and that viruses are actually obligate parasites. The quantitative relationships can then be investigated with confidence and one should be able to determine conclusively whether or not the inclusion-type viruses survive in their hosts in an inapparent state.

Quantitative Virus-Host Relationships

The inclusion-type viruses appear to depend largely upon mechanical contamination of the food of their hosts for their continued dissemination and survival. As stated previously, they are well-suited to survival outside their hosts. The virus from a single diseased larva may be spread over the food surface and remain viable for a considerable period of time causing the infection of a large number of hosts (CLARK, 1958; BIRD, 1961). Experimental, quantitative studies have revealed important information concerning the numerical relationships between the viruses and their host populations (BENZ, 1964; STAIRS, 1964, 1965).

Viruses are relatively easy to obtain in large quantities and in a pure state. A diseased larva can be macerated in water and fed to a large number of hosts which become infected and die yielding more virus. This can be repeated until sufficient quantities for the proposed studies are obtained. Inclusion bodies are resistant to putrefaction so the hosts can be suspended in water and allowed to disintegrate. The inclusion bodies may then be obtained by differential centrifugation because they are heavier than any of the other particles of equal size in the diseased larvae. Repeated centrifugation will provide a very pure and uniform suspension of inclusion bodies. The concentration of suspensions can be determined in a number of ways; but since the inclusions are large enough to be seen under the light microscope, the simplest and most direct method is the use of a standard bacteria counting chamber. The new electric counters, however, may prove to be more convenient and more accurate.

Controlled quantitative tests are very dependent upon disease-free hosts. In most cases, a disease-free host population can be obtained either from natural field populations or by intensive selective breeding and controlled rearing (BIRD, 1949; STAIRS, 1963, 1965). Once a stock is obtained, it is very important to keep it free from disease. This is done by rearing larvae individually from the egg on virus-free food in sterile containers. The development of artificial diets has helped to eliminate the variable of contaminated food. Efforts should also be made to control the genetics of the host populations and the first quantitative tests should be performed using a sample of the population with the widest genetic variability (STAIRS, 1965). Subsequently, specific host lines can be selected and studied more effectively and realisticaly.

Measured quantities of virus may be fed to test larvae or injected into the body cavity. Before injections are successful, the inclusion bodies must be dissolved in weak alkali (0.05 M—0.005 M Na_2CO_3 + 0.05 NaCl solutions) to release the virions. The released virions may be concentrated by centrifugation at

12,000 g or higher (BERGOLD, 1947). The inclusions of granulosis virus may be treated in the same manner to release the virions but a slightly higher concentration of sodium carbonate (0.1 M—0.03 M) is often required for effective results. The cytoplasmic polyhedra behave differently in weak alkaline solutions. They are more difficult to dissolve but treatment with 0.1 M Na_2CO_3 or 0.2 M NaOH is effective (KAWASE, 1964; HOSAKA and AIZAWA, 1964). Virions released by these methods have been examined under the electron microscope and viability tests have shown that they retain their infectivity for midgut epithelium. These spherical virions are relatively smaller (69 mμ) than the rod-shaped virions and they require centrifugation at 50,000 to 100,000 g for adequate concentration.

Quantitative studies on the infection process of a granulosis virus of the larch budmoth, *Zeiraphera diniana* (GUÉNÉE), indicate that once the virus reaches susceptible tissues multiplication proceeds in an apparently defined way (BENZ, 1964). A single virion can infect a fourth-instar larva and cause death in about 13 days. Similarly, 10^4 virions per larva appears to be sufficient for the occupation of all the sites of virus multiplication in one larva causing death in the minimum time of four days. Much of the virus fed to these larvae does not reach susceptible tissues. An average estimate indicates that only one particle in 10^3 or 10^4 reaches susceptible tissues. In other words, 99.9 to 99.99% of the virus consumed by these larvae is inactivated or excreted and does not succeed in causing infection. These studies have shown that the quantitative approach may be very essential to understanding the virus infection process.

Using the quantitative approach, it has been possible to obtain information essential to understanding the development of epizootics and the subsequent manipulation of viruses for successful pest management. For example, all the larvae in a given population are not equally susceptible to virus infection. Susceptibility may vary with the age of larvae but even at a given age there is often wide variation. In a group of first-instar *M. disstria* larvae some individuals may be susceptible to a single polyhedron, whereas, others may be resistant to a dosage of 10^3 polyhedra per larva (STAIRS, 1965). These differences are quite large and they indicate considerable variation in susceptibility even in the very early larval stage. As larvae grow older they may become increasingly resistant. Fourth-instar *M. disstria* larvae may be tolerant of dosages of 10^9 polyhedra per larva. This demonstrates that hosts are able to resist infection by virus in varying degrees and that resistance factors may increase with larval age. The mechanism of this resistance is unknown but a systematic investigation would probably reveal the lumen of the migdut plays a major role (AIZAWA, 1962). This assumption is based on studies of a different virus-host system (nuclear-polyhedrosis of *G. mellonella*) where it was estimated that at LD_{50} dosages, 97% of the virus ingested was inactivated in the midgut and none of the virus appeared to be excreted (STAIRS, 1963). Many qualitative observations suggest that an increase of virus resistance with increasing larval age may be common among the Lepidoptera (TANADA, 1953; MORRIS, 1962; CLARK and REINER, 1956).

Similar data also indicate that there may be exceptions to this generalization. For example, the cabbage looper, *Trichoplusia ni* (HUBNER) appears to have a relatively narrow range of susceptibility to a nuclear-polyhedrosis virus during the entire larval stage (JAQUES, 1961). By the same token, the sawflies show a similar narrow range of response to their nuclear-polyhedrosis viruses throughout the larval stage (BIRD and WHALEN, 1953; KRIEG, 1955).

A knowledge of quantitative virus-host relationships is important when one considers the susceptibility of a host to a number of different viruses. The nuclear-polyhedrosis virus of *B. mori* is infectious for *G. mellonella* (AIZAWA, 1962) but the dosage required to cause infection in the latter is much greater. The present author found that one *G. mellonella* LD_{50} unit is equal to approximately 40,000 *B. mori* LD_{50} units and the virus develops at a slower rate in the unnatural host. These results indicate the nuclear-polyhedrosis virus from one host may be capable of developing in a number of different hosts provided that a high enough concentration is administered to the unnatural hosts. Using this approach, it may be possible to adapt viruses to economically important species that are not presently susceptible to a virus.

Certain species of Lepidoptera are known to be susceptible to viruses normally infecting closely related species. Quantitative studies, for example, have shown that the nuclear-polyhedrosis viruses from North American species of the genus *Malacosoma* are identical as far as their virulence for *M. disstria* is concerned (STAIRS, 1964). A virus from a European species is also infectious for *M. disstria* but appears to be less virulent and slower acting. This type of information is valuable when one considers the spread and survival of these viruses in their host populations. It is conceivable that a virus would be better able to survive as a species if it were capable of infecting a number of closely-related host species. Natural dispersion of a virus from one sympatric host species to another may account for some of the sudden epizootics observed in certain species of Lepidoptera.

Quantitative studies have also shown that granulosis viruses from two races of *Pieris brassicae* L. are of equal virulence but the host races themselves are relatively different in their response to both viruses (DAVID and GARDINER, 1965). One race is much more resistant than the other. This suggests that in this virus-host sytem the population response of the host is more elastic than virus virulence. Perhaps maximum virulence has evolved without completely destroying the host species. This also appears to be true for the nuclear-polyhedrosis viruses of *Trichoplusia ni* (HUBNER) and *Diprion hercyniae* (HTG.).

The silkworm, *Bombyx mori* (L.), shows a variable response to a cytoplasmic polyhedrosis virus depending on the genetic constitution of the test strain (ARUGA and WATANABE, 1964). Breeding experiments indicate that resistance may be controlled by a dominant major gene (WATANABE, 1965). A median lethal concentration for the resistant strain was 10^9 polyhedra per ml but for the susceptible strain it was $10^{6.5}$ polyhedra per ml. When compared to the other inclusion-type viruses both values are relatively high but they do indicate that one strain is about 300 times more resistant than the other. Morphological differences in the cytoplasmic polyhedrosis viruses are known but they appear to be equally virulent (ARUGA and WATANABE, 1964).

Quantitative studies on a granulosis virus of *Zeiraphera diniana* (GUÉNÉE) have shown that geographically separated populations respond differently to the virus (BENZ, 1962, 1964). Larvae from a sparse, endemic population (Engadin, Switzerland) were twice as susceptible as those from the more severe infestations (BRIANÇONNAIS, FRANCE and MONTGENEVRE, Italy) when the 50% lethal dosages were compared. When the 25 % lethal dosages were compared, however, individuals from the sparse population were five or six times as susceptible as those from the dense population. It is apparent that the sparse population is more susceptible to lower dosages but less susceptible to higher dosages and therefore, more variable in response to virus. Since the same virus was used in the tests, the genetic constitution of the host populations in the heavily infested areas must have been different from that in the sparsely populated areas.

Similar results have been obtained in studies of population response to nuclear-polyhedrosis viruses. Geographically separated populations of the California oakworm, *Phryganidia californica* PACKARD, respond differently to a nuclear-polyhedrosis virus in a manner that appears to be very similar to that of *Zeiraphera diniana* (GUÉNÉE) (MARTIGNONI and SCHMID, 1961). The nuclear-polyhedrosis virus of *Heliothis* species appears to cause a different response among larvae of *H. zea* and *H. virescens* (IGNOFFO, 1966). At the lower dosages, *H. virescens* larvae are relatively more susceptible but at higher dosages they are more resistant than *H. zea*. This indicates more variability among individuals of *H. virescens* than among those of *H. zea*.

These studies correlate well with those on the granulosis virus of *Pieris brassicae* where the virus from widely separated geographical areas appeared to be similar but the host larvae responded differently (DAVID and GARDINER, 1965). Apparently, these viruses have evolved maximum virulence and the variation observed in the virus-host interaction is largely the result of the variability of host response to virus infection.

Natural Occurence and Dispersal

Inclusion-type viruses have been isolated from more than 300 species of insects representing the orders Lepidoptera, Hymenoptera, Diptera, Orthoptera, and Neuroptera (AIZAWA, 1963; SMITH, 1963; HUGER, 1963). The best known are the nuclear-polyhedrosis and granulosis groups and these are largely confined to the Lepidoptera and Hymenoptera (SAWFLIES). Most species of sawflies in the genera *Diprion* and *Neodiprion* are susceptible only to nuclear-polyhedrosis viruses that attack the digestive epithelium of the midgut. Many species of Lepidoptera are susceptible to more than one inclusion-type virus (TANADA, 1956; STEINHAUS, 1957; BIRD, 1959). The cytoplasmic polyhedrosis viruses appear to be very widespread among this group. They often occur simultaneously in the same host with the nuclear-polyhedrosis and granulosis viruses, and it is quite probable that many so called nuclear-polyhedrosis virus suspensions are mixtures of the two types of virus.

The nuclear-polyhedrosis viruses are found primarily in species of the families Arctiidae, Bombycidae, Geometridae, Lasiocampidae, Lymantridae, Nymphalidae, Olethreutidae, Phalaenidae (Noctuidae), Pieridae, Psychidae, Thaumetopoeidae, Tineidae, and Tortricidae (HUGHES, 1957; MARTIGNONI and LANGSTON, 1960). The granulosis viruses are often found in species of many of the above families but most frequently in those of Phalaenidae, Pieridae, Olethreutidae, and Tortricidae. Granulosis viruses have not been found in Bombycidae, Lasiocampidae, and Lymantridae. Single species of the Phalaenidae, Olethreutidae and Tortricidae often are attacked by both granulosis and nuclear-polyhedrosis viruses. Thus, complicated virus interactions may be encountered when studying natural populations of these species. By the same token, many of these species are susceptible to various bacteria, protozoa and fungi which may also interact with the viruses.

The protozoa are probably the most common pathogens occuring in populations in association with viruses. They usually cause debilitative diseases resulting in reduced viability of the host, whereas, inclusion-type viruses consistently cause lethal diseases. Because of this, many individuals that contract a virus infection may have been previously infected with a protozoan and at death contain both pathogens. The protozoa do not seem to interfere with the virus infection because the hosts infected with protozoa appear to be equally susceptible to infection by virus. A virus usually develops much faster than protozoa and is capable of multiplying in tissues that are already supporting protozoa. On the other hand, if bacteria and fungi gain access to tissues that have been prevoiusly infected by a virus, these organisms usually destroy the tissues before the virus can complete its development. Their presence seems to be detrimental to the survival of virus, whereas, the presence of protozoa does not appear to inhibit virus survival.

Multiple virus infections have been studied and it is evident that different viruses interact. If larvae of *Choristoneura fumiferana* (CLEMENS) are infected simultaneously with nuclear-polyhedrosis and granulosis viruses they develop nuclear-polyhedrosis but if they are fed granulosis virus two days before they are fed nuclear-polyhedrosis virus most of them become infected with granulosis virus (BIRD, 1959). The granulosis virus appears to be slower acting and requires a time advantage to compete successfully with the nuclear-polyhedrosis virus. Although both viruses attack the same tissues of the host, both are not found infecting the same cell. Adjacent cells often are infected with different viruses. This indicates a cell that has received a specific virus is resistant to, or incapable of, receiving the other virus. Apparently, these viruses interfere with one another at the cellular level but not at the tissue level. Either virus given a temporal advantage is capable of occupying most of the susceptible cell sites.

This does not appear to be the case when *Pseudaletia unipuncta* larvae are fed a similar combination of viruses. The presence of the granulosis virus appears to cause the host larvae to be more susceptible to the nuclear-polyhedrosis but the opposite effect does not occur (TANADA, 1959). Heat-inactivated granulosis virus has a similar effect. This has been referred to as a synergistic reaction between two viruses. It is possible, however, that the host is made more susceptible because its defense mechanisms are occupied by the granulosis virus thus giving the nuclear-polyhedrosis virus more opportunity to reach susceptible cells. The susceptibility of *G. mellonella* larvae to virus increased when their phagocytes were blocked with carbon particles (STAIRS, 1964). This was clearly not synergism but decreased host susceptibility.

The nuclear-inclusion-type viruses have often been observed to cause the decimation of their host populations, particularly, at periods of high host density (ESCHERICH, 1913; GLASER, 1915; BALCH and BIRD, 1944; STEINHAUS, 1949). Epizootics usually develop very rapidly and are quite spectacular in that so many host larvae are killed by the virus. Within a generation or two the host population is greatly reduced. A single virus-type is usually involved instead of a number of different viruses. In most instances, a great many larvae appear to suddenly become diseased and mortality is very evident. This usually occurs after the host populations have maintained a high level for a number of generations and superficial observations would suggest that an environmental change causes a physiological change in the host that allows, or triggers, virus development. Such a theory assumes that these viruses are capable of spreading through the host population in somewhat the same manner as host genes without causing any cytopathic cellular changes or mortality. The virus population increases with the host population and could, therefore, survive indefinitely without ever causing the death of its host. This is a complicated theory that has not been substantiated at all by comprehensive laboratory and field studies.

Perhaps the most important study made on a natural virus-host system is the nuclear-polyhedrosis virus of the European spruce sawfly, *Diprion hercyniae* (HARTIG). The host was accidently introduced into the North American continent from Europe and began to reach exceedingly high population levels in the 1930's that threatened the extensive spruce forests in the east. Entomologists tried to combat this threat by importing numerous natural parasites from Europe and releasing them into the North American populations. During this program, the virus, which occurs naturally in European populations, was accidently introduced. The effect of the virus was spectacular. Host densities were reduced rapidly, the virus spread quickly throughout the area and many valuable acres of spruce forests were saved (BALCH and BIRD, 1944).

This virus-host system has been monitored continuously in field populations and studied intensively in the laboratory since 1942 (BIRD and ELGEE, 1957). These studies have demonstrated that the virus is extremely contagious and virulent to all larval stages but the embryonic regenerative cells are resistant to infection (BIRD, 1953). Thus, individuals that contact virus very late in the larval stage often undergo metamorphsis to the adult stage where the embryonic cells become functional and susceptible. The infected adults lay a full compliment of eggs and transmit virus to a part of their progeny which becomes infected and dies during the early larval instars.

With careful sterile techniques, the host can be reared disease-free in the laboratory, but, as soon as precautions are relaxed, the virus spreads through the host populations again. Subsequent careful individual rearing of larvae always results in a disease-free colony even though the starting culture may have been heavily diseased. These results show that the virus must depend largely upon simple contamination of food for its survival in the host population. It appears to be egg-transmitted for only one generation and only those individuals that feed on contaminated food just before pupating transmit the virus to their progeny. Intensive investigation of the mode of egg transmission will probably show that eggs are, essentially, mechanically contaminated and that the degree of transmission depends on the level of infection of the adult.

The nuclear-polyhedrosis viruses of other sawfly species and those of certain Lepidoptera exhibit many of the same attributes. Those of *N. sertifer*, *N. lecontei*, and *N. swainei* are very similar to the *D. hercyniae* virus in their association with their hosts (BIRD, and WHALEN, 1953; SMIRNOFF, 1961). The viruses of species in the lepidopteran genus *Malacosoma* also appear to depend on mechanical dispersion for their continued survival (STAIRS, 1966). Disease-free populations of *M. disstria* can be reared in the laboratory provided the larvae are kept individually in separate sterile containers. The virus is so contagious that it is nearly impossible to keep large rearing containers, benches, or rooms decontaminated. First-instar larvae are very susceptible and some can be infected by a single polyhedron. As the larvae grow older they become increasingly resistant to the virus and in the last instar many larvae can consume relatively large quantities of viruses without becoming infected. Adults transmit virus to their progeny and histopathological studies show that both sexes develop overt infections. This indicates the virus may be transmitted by adults of either sex.

Experimental field studies of the nuclear-polyhedrosis virus of *M. disstria* have shown that adult transmission is very important to the development of widespread epizootics (STAIRS, 1965). In a single generation virus was spread 30 miles from the point of introduction and more or less evenly into the whole host population. Much of the dispersal was into areas not previously infected by the host, further demonstrating the effectiveness of adult transmission.

Calculations, based on quantitative studies, show the virus is capable of survival and spread among dense larval populations as the result of simple mechanical contamination of foliage. A single, second-instar larva dying from virus received from one of its parents contains about 10^7 polyhedra. This is enough virus to infect at least 5,000 second and third instar larvae and, since these larvae feed in large compact colonies, many of them will most certainly become infected by eating foliage contaminated by this larva. As long as host-density is relatively high the virus will survive and spread. Dispersion, however, appears to be assisted by almost anything that moves through the host population. The rain, wind, parasites, predators, scavengers all appear to be important (BIRD, 1961; FRANZ, 1956). The action of rain is confined to spreading the virus down the host plant and the wind may carry infected early-instar larvae short distances. Scavengers and predators spread viruses from plant to plant. Specific parasites may be very important in the dispersal of virus among larval populations of *M. disstria*. One dipterous parasite, *Sarcophaga aldrichi* (PARKER), appears to be essential to the rapid development of epizootics (STAIRS, 1966). Adults of this species are attracted to the recently dead, virus-diseased larvae on which they feed and over which they crawl. Hence, they become highly contaminated with fresh virus. When they are not feeding much time is spent searching over the foliage for prospective hosts. During this search widespread contamination of the foliage occurs.

The rapid development of epizootics in *M. disstria* populations seems to depend on the presence of numerous foci of infection and a relatively high population of active *S. aldrichi* adults. In populations where there are numerous foci of adult-transmitted virus and a low population of parasites, epizootics develop slowly, but where there are relatively few foci and numerous parasites, epizootics develop more rapidly and reach a higher intensity. This indicates the parasite may be more important to the development of large-scale epizootics than a large number of foci.

The rate at which a virus epizootic develops in a host population is also believed to be influenced considerably by temperature. Epizootics develop more slowly at suboptimum temperatures (BIRD, 1955; SMIRNOFF, 1967). The temperature of the environment appears to affect the development of virus disease and, in most cases, the optimum temperature for host development is probably the best also for virus development. Lower temperatures slow the development of hosts and virus, whereas, higher temperatures tend to cause an increase in the rate of both host and virus development. In some cases, a high temperature will inhibit virus development. The larvae of *Diprion hercyniae* (HARTIG) are not susceptible to nuclear-polyhedrosis virus when hosts are kept at a constant temperature of 29.4° C (BIRD, 1955). Similarly, larvae of the *Trichoplusia ni* (HUBNER) and *Heliothis zea* (BODDIE) are resistant to nuclear-polyhedrosis virus when reared at 36° C but *Prodenia sp.* larvae are susceptible at temperatures up to 45° C (THOMPSON, 1959). It appears that high temperature causes certain hosts to become resistant while others do not seem to be changed. The importance of high temperature to the development of epizootics has not been investigated but in view of these laboratory studies its effect should be considered.

References

AIZAWA, K.: Mode of multiplication of silkworm nuclear polyhedrosis virus. II. Multiplication of the virus in the early period of the LD_{50} time curve. J. Insect Pathol. **1**, 67—74 (1959).

— Antiviral substance in the gut juice of the silkworm, *Bombyx mori* L. J. Insect Pathol. **4**, 72—76 (1962).

— Infection of the greater wax moth, *Galleria mellonella* (Linnaeus) with the nuclear polyhedrosis virus of the silkworm, *Bombyx mori* (Linnaeus). J. Insect Pathol. **4**, 122—127 (1962).

AIZAWA, K.: The nature of infections causes by nuclear-polyhedrosis viruses. In: Insect pathology, an advanced treatise, (ed. E. A. STEINHAUS), vol. 1, p. 381—412. New York: Academic Press 1963.
—, and S. IIDA: Nucleic acids extracted from the virus polyhedra of the silkworm, *Bombyx mori* (Linnaeus). J. Insect Pathol. **5**, 344—348 (1963).
ARUGA, H.: Induction of virus infections. In: Insect pathology, an advanced treatise (ed. E. A. STEINHAUS), vol. 1, p. 499—530. New York: Academic Press 1963.
—, and E. NAGASHIMA: Generation to generation transmission of cytoplasmic polyhedrosis virus of *Bombyx mori* (Linnaeus). J. Insect Pathol. **4**, 313—320 (1962).
—, and H. WATANABE: Resistance to *per os* infection with cytoplasmic-polyhedrosis virus in the silkworm, *Bombyx mori* (Linnaeus). J. Insect Pathol. **6**, 387—394 (1964).
BALCH, R. E., and F. T. BIRD: A disease of the European spruce sawfly *Gilpinia hercyniae* (HTG.) and its place in natural control. Sci. Agric. **25**, 65—80 (1944).
BENZ, G.: Untersuchungen über die Pathogenität eines Granulosis Virus des Grauen Larchenwicklers *Zeiraphera diniana* (GUÉNÉE). Agronomski glasnik (Zagreb) 566—573 (1962).
— Aspects of virus multiplication and average reduplication time for a granulosis virus of *Zeiraphera diniana* (GUÉNÉE). Entomophaga Memoire No 2, 418—421 (1964).
BERGOLD, G. H.: Über Polyederkrankheiten bei Insekten. Biol. Zbl. **63**, 1—55 (1943).
— Die Isolierung des Polyedervirus und die Natur der Polyeder. Z. Naturforsch. **26**, 122—143 (1947).
— Viruses of insects. In: Handbuch der Virusforschung (Hrsg. C. HALLAUER and K. F. MEYER), Bd. 4, S. 60—142. Wien: Springer 1958.
— The nature of nuclear-polyhedrosis viruses. In: Insect pathology, an advanced treatise (ed. E. A. STEINHAUS), vol. 1, p. 413—456. New York: Academic Press 1963.
—, and J. SUTER: On the structure of cytoplasmic polyhedra of some Lepidoptera. J. Insect Pathol. **1**, 1—14 (1959).
BIRD, F. T.: A virus (polyhedra) disease of the European spruce sawfly *Gilpinia hercyniae* (HTG.). Thesis, McGill University, Montreal, Canada 1949.
— The effect of metamorphosis on the multiplication of an insect virus. Canad. J. Zool. **31**, 300—303 (1953).
— Virus diseases of sawflies. Canad. Entomol. **87**, 124—129 (1955).
— Polyhedrosis and granulosis viruses causing single and double infections in the spruce budworm, *Choristoneura fumiferana* (CLEMENS). J. Insect Pathol. **1**, 406—430 (1959).
— Transmission of some insect viruses with particular reference to ovarial transmission and its importance in the development of epizootics. J. Insect Pathol. **3**, 352—380 (1961).
— On the development of insect polyhedrosis and granulosis virus particles. Canad. J. Microbiol. **10**, 49—52 (1964).
— The development of a cytoplasmic polyhedrosis virus in the spruce budworm, *Choristoneura fumiferana* (CLEMENS). Canad. J. Microbiol. **12**, 337—339 (1966).
—, and D. E. ELGEE: Virus disease and introduced parasites as factors controlling the European spruce sawfly, *Diprion hercyniae* (HTG.) in central New Brunswick. Canad. Entomol. **89**, 371—378 (1957).
—, and M. M. WHALEN: Stages in the development of two insect viruses. Canad. J. Microbiol. **1**, 170—174 (1954).
BOLLE, J.: Die Gelb- oder Fettsucht der Seidenraupe, eine parasitäre Krankheit. In: Der Seidenbau in Japan, S. 94—141. Budapest: Hartlebens 1898.
CLARK, E. C.: Ecology of the polyhedroses of tent caterpillars. Ecology **39**, 132—139 (1958).
—, and C. E. REINER: The possible use of a polyhedrosis virus in the control of the great basin caterpillar. J. econ. Entomol. **49**, 653—659 (1956).

DAVID, W. A. L., and B. O. C. GARDINER: Resistance of *Pieris brassicae* (Linnaeus) to granulosis virus and the virulence of the virus from different host races. J. Invert. Pathol. **7**, 285—290 (1965).
ESCHERICH, K.: Neues über Polyederkrankheiten. Naturw. Z. Land- u. Forstwirtsch. **11**, 86—97 (1913).
ESTES, Z. E., and R. M. FAUST: Silicon content of intact nuclear polyhedra from the corn earworm, *Heliothis zea*. J. Invert. Pathol. **8**, 145—159 (1966).
FAULKNER, P.: Isolation and analysis of ribonucleic acid from inclusion bodies of the nuclear polyhedrosis of the silkworm. Virology **16**, 479—484 (1962).
FAUST, R. M., and J. R. ADAMS: The silicon content of nuclear and cytoplasmic viral inclusion bodies causing polyhedrosis in Lepidoptera. J. Invert. Pathol. **8**, 526—530 (1966).
FRANZ, J.: Die künstliche Verbreitung von Virosen einiger Blattwespen (Diprionidae) innerhalb und außerhalb ihres Endemiegebietes. Verh. dtsch. Ges. Zool. Er. 1955, 407—412. (1956).
GLASER, R. W.: Wilt of gypsy moth caterpillars. J. agric. Res. **4**, 101—128 (1915).
HOSAKA, Y., and K. AIZAWA: The fine structure of the cytoplasmic-polyhedrosis virus of the silkworm, *Bombyx mori* (Linnaeus). J. Insect Pathol. **6**, 53—77 (1964).
HUGER, A.: Über die Natur des Fadenwerkes bei der Granulose von *Choristoneura murinana* (HBR.) (Lepidoptera, Tortricidae). Naturwissenschaften **47**, 358—359 (1960).
— Granulosis of insects. In: Insect pathology, an advanced treatise, (ed. E. A. STEINHAUS), vol. 1, p. 531—575. New York: Academic Press 1963.
HUGHES, K. M.: An annotated list and bibliography of insects reported to have virus disease. Hilgardia (Berkeley, Calif.) **26**, 597—629 (1957).
HUKUHARA, T., and Y. HASHIMOTO: Serological studies of the cytoplasmic and polyhedrosis viruses of the silkworm, *Bombyx mori*. J. Invert. Pathol. **8**, 234—239 (1966).
IGNOFFO, C. M.: Susceptibility of the first instar of the bollworm, *Heliothis zea* and the tobacco budworm, *Heliothis virescens* to *Heliothis* nuclear-polyhedrosis virus. J. Invert. Pathol. **8**, 531—536 (1966).
JAQUES, R. P.: The influence of physical stress on growth and nuclear polyhedrosis of *Trichoplusia ni* (HUBNER). J. Insect Pathol. **3**, 47—54 (1961).
KAWASE, S.: The amino-acid composition of viruses and their polyhedron proteins of the polyhedroses of the silkworm, *Bombyx mori* (Linnaeus). J. Insect Pathol. **6**, 156—163 (1964).
—, and H. YOSHIYUKI: Nucleic acid and protein changes in blood and midgut of the silkworm, *Bombyx mori* (Linnaeus), during the course of cytoplasmic polyhedrosis. J. Invert. Pathol. **7**, 49—54 (1965).
KOMANO, T., M. HIMENO, Y. OHNO, and K. ANODERA: Studies on lipids in the haemolymph of the silkworm, *Bombyx mori*, during the course of nucleopolyhedrosis. J. Invert. Pathol. **8**, 67—74 (1966).
KRIEG, A.: Untersuchungen über die Polyhedrose von *Neodiprion sertifer* (GEOFFR.) Arch. ges. Virusforsch. **6**, 163—174 (1955).
KRYWIENCZYK, J.: Demonstration of nuclear-polyhedrosis in *Bombyx mori* (Linnaeus) by fluorescent antibody technique. J. Insect Pathol. **5**, 309—317 (1963).
—, and G. H. BERGOLD: Serological relationships between inclusion body proteins of some Lepidoptera and Hymenoptera. J. Immunol. **84**, 404—408 (1960).
— — Serological relationships between insect viruses and their inclusion-body proteins. J. Insect Pathol. **2**, 118—123 (1960).
— — Serological relationships of viruses from lepidopterous and hymenopterous insects. Virology **10**, 308—315 (1960).
MARTIGNONI, M.E., and R.L. LANGSTON: Supplement to an annotated list and bibliography of insects reported to have virus diseases. Hilgardia (Berkeley, Calif.) **30**, 1—40 (1960).

MARTIGNONI, M. E., and J. E. MILSTEAD: Hypoproteinemia in a noctuid larva during the course of a nucleopolyhedrosis. J. Insect Pathol. **6**, 517—531 (1964)
—, and P. SCHMID: Studies on the resistance to virus infections in natural populations of Lepidoptera. J. Insect Pathol. **3**, 62—74 (1961).
MORRIS, O. N.: Quantitative infectivity studies on the nuclear polyhedrosis virus of the western oak looper, *Lambdina fiscellaria somniaria* (HULST). J. Insect Pathol. **4**, 207—215 (1962).
NEILSON, M. M.: Effects of a cytoplasmic polyhedrosis on adult Lepidoptera. J. Invert. Pathol. **7**, 306—314 (1965).
PAILLOT, A.: Existence de la grasserie chez les papillons de ver a soie. C. R. Acad. Agr. **12**, 57—63 (1926).
SMIRNOFF, W. A.: A virus disease of *Neodiprion swainei* MIDD. J. Insect Pathol. **3**, 29—46 (1961).
— Influence of temperature on the development of a virus disease of *Neodiprion swainei* (Hymenoptera, Tenthredinidae). Canad. Entomol. **99**, 236—243 (1967).
SMITH, K. M.: The cytoplasmic virus diseases. In: Insect pathology, an advanced treatise, (ed. E. A. STEINHAUS), vol. 1, p. 457—497. New York: Academic Press 1963.
STAIRS, G. R.: A study of virus infection in Lepidoptera with special reference to the nuclear-polyhedrosis of *Galleria mellonella* L. Thesis, McGill University (1963).
— Infection of *Malacosoma disstria* (HBR.) with nuclear-polyhedrosis viruses from other species of *Malacosoma* (Lepidoptera, Lasiocampidae). J. Insect Pathol. **6**, 164—169 (1964).
— Changes in the susceptibility of *Galleria mellonella* (Linnaeus) larvae to nuclear-polyhedrosis virus following blockage of the phagocytes with India ink. J. Insect. Pathol. **6**, 373—376 (1964).
— Selection of a strain of insect granulosis virus producing only cuboidal inclusion bodies. Virology **24**, 520—521 (1964).
— Quantitative differences in susceptibility to nuclear-polyhedrosis virus among larval instars of the forest tent caterpillar, *Malacosoma disstria* (HUBNER). J. Invert. Pathol. **7**, 427—429 (1965).
— Dosage-mortality response of *Galleria mellonella* L. to a nuclear-polyhedrosis virus. J. Invert. Pathol. **7**, 5—9 (1965).
— The effect of metamorphosis on nuclear-polyhedrosis virus infection in certain Lepidoptera. Canad. J. Microbiol. **11**, 509—512 (1965).
— Transmission of virus in forest tent caterpillar populations. Canad. Entomol. **98**, 110—114 (1966).
STEINHAUS, E. A.: Principles of insect pathology. New York: McGraw-Hill Book Co. 1949.
— New records of insect virus diseases. Hilgardia (Berkeley, Calif.) **26**, 417—430 (1957).
— Crowding as a possible factor in insect disease. Ecology **39**, 503—514 (1958).
TANADA, Y.: Description and characteristics of a granulosis virus of the imported cabbage worm. Proc. Hawaiian Entomol. Soc. **15**, 235—260 (1953).
— Some factors affecting the susceptibility of the armyworm to virus infections. J. econ. Entomol. **49**, 52—57 (1956).
THOMPSON, C. G.: Thermal inhibition of certain polyhedrosis virus diseases. J. Insect Pathol. **1**, 89—90 (1959).
WATANABE, H.: Resistance to peroral infection by the cytoplasmic-polyhedrosis virus in the silkworm, *Bombyx mori* (Linnaeus). J. Invert. Pathol. **7**, 257—258 (1965).
WELLINGTON, E. F.: The amino acid composition of some insect viruses and their characteristic inclusion-body proteins. Biochem. J. **57**, 334—338 (1954).
WYATT, G. R.: The nucleic acids of some insect viruses. J. gen. Physiol. **36**, 201—205 (1952).
YAMAFUGI, K., and F. HASHINAGA: Isolation of a pre-viral deoxyribonucleic acid. Nature (Lond.) **201**, 1158—1159 (1966).

Non-Inclusion Virus Diseases of Invertebrates

C. VAGO*

Introduction

In virology, as in physics, chemistry or other branches of science, there occur all over the world certain periods during which research concentrates on the same type of subject. Such trends are a sort of fashion in science. These waves of fashion often spring from profound reasons, based on the perfection of new techniques that open new lines of research. Subsequently, important results open new perspectives during longer or shorter periods.

Such a new dynamic trend is now apparent in invertebrate virology. Indeed, for over a century, the only virus diseases known among invertebrates were the polyhedroses. The reason for this exclusiveness was not only economic — a serious illness of silkworms. Above all, the polyhedral inclusion bodies were extremely easy to recognize thanks to their large size, typical shape, their brilliance, and their abundance. Later, for the same reasons, other types of viruses were discovered, owing to the easy demonstration of inclusion bodies of varying shapes.

Thus, only a few years ago, the general opinion in patholgy was that virus diseases of invertebrates were necessarily linked with the formation of inclusion bodies and only seven years ago it would have been inconceivable to write a chapter on invertebrate virus diseases devoid of inclusion bodies.

Presently, invertebrate virologists increasingly employ in their virus studies methods of investigation which have proved themselves in medical and veterinary virology. As a result, in the last dozen years more than thirty virus diseases without inclusion bodies have been described in invertebrates.

Here will be presented first the various diseases, grouping them according to type, then evaluating their meaning in compared pathology and finally discussing them from the point of view of applied virology.

Types of Viroses Described

Viroses in which free bodies have been recognized, display on the one hand strong resemblances and, on the other hand, essential differences in the shape and pathogenesis of the viruses.

* Institute for Cytopathology. St. Christol-les-Ales, Gard, France.

Sacbrood of the Bee

This is the insect disease whose viral nature has been established for the longest time. From the beginning of this century cases of foulbrood have been found among bee larvae, in which no bacteria could be detected even after death. It is probable that the disease in question is the one to which WHITE gave the name of sacbrood in 1913 and which differs from typical foulbroods by the absence of putrefaction and the formation of serosity in the epidermis. The infectivity of the filtrate of diseased larvae was demonstrated (WHITE, 1917) and later confirmed (MORGENTHALER, 1947) but no inclusion bodies were noted by the authors. In the first attempt to classify insect viruses (HOLMES, 1948) the genus *Morator* was created to distinguish this type of disease from other insect viruses which formed inclusion bodies. Under the electron microscope, round particles of 60 mμ were at first noticed (STEINHAUS, 1949), then paraspherical viruses of 28 mμ (VAGO, 1964), of 30 mμ (BŘČAK, SVOBODA and KRÁLÍK, 1963; BŘČAK and KRÁLÍK, 1965) and of 28 mμ (BAILEY, GIBBS and WOODS, 1964), were described. These latter authors evaluated the LD_{50} for 2-day-old larvae as 10^5—10^6 particles, and showed the serological identity of the different isolated strains. Similar particles have been recognized in ultrathin sections of non-defined tissue (LEE and FURGALA, 1965 a). The nucleic acid was defined as RNA (LEE and FURGALA, 1965 b) but no information on its pathogenesis has been published.

Viral Disease of *Cirphis unipuncta* (HAWORTH)

Spherical or oval particles of 25 mμ have been noted in sick larvae, of a darker color than healthy insects, and containing neither bacteria nor inclusion bodies (STEINHAUS, 1951; WASSER, 1952). However, no further studies have been made on the virus structure and pathogenesis.

Transparency of Scarabeidae

Another case of a paraspherical virus, freely scattered in the cytoplasm has been noted in an illness of *Scarabeidae* known for a long time. HEIDENREICH (1939) described a state called „Wassersucht", consisting of the disappearance of the adipose tissue and the pronounced transparence of the whole body of *Melolontha melolontha* L. larvae. This author's transmission experiments permitted to envisage a viral etiology which was confirmed by the observation of icosahedral-like particles of 60 mμ to 75 mμ in the fat body (KRIEG and HUGER, 1960). The cytoplasm was replaced by negative Feulgen strands and after the destruction of the cells the virus remained in the cytoplasm or dispersed in the haemolymph. The particles containing RNA have been named *Moratorvirus lamellicornium*.

Paralysis of Crickets

The first viral disease among Orthoptera was discovered in breeding crickets, *Gryllus bimaculatus* GEER. Diseased insects show symptoms of paralysis and trembling of the legs as well as an inability to move (MEYNADIER,

1966). An icosahedral virus of about 30 mμ in diameter was isolated and the disease was reproduced by inoculation of the purified suspension of this virus. Studies are under way to determine the affinity of this virus to the nervous tissue of crickets.

Paralysis of Crabs

This is the first virosis of invertebrates other than insects and arachnids. A progressive paralysis has been observed at several occasions among crabs *Macropipus depurator* L. on the Mediteranean coast (VAGO, 1966b). Paraspherical viral particles of 50—60 mμ have been isolated from muscles and glandular tissue. These particles were infectious when inoculated into crabs.

The observation of corpuscles of the same shape and diameter in the androgene gland of the crab *Carcinus maenas* L. suffering from a similar illness indicated that viral diseases of Crustacea may be more common than suspected earlier.

Flaccidity in Silkworms

One of the oldest problems in insect pathology is that of intestinal diseases which have been noted for more than 200 years in silkworm rearing. Their etiology has often given rise to violent controversy among authors that studied the diseases for economic reasons. We know today that several etiological types exist under the "flaccidity" denomination introduced by PASTEUR (1868). Some are of a physiological nature while others are due to the action of *Bacillus thuringiensis*, var. *sotto* and *alesti*. On the basis of transmission experiments, one type, called the "clear heads disease", was considered by PAILLOT (1930a, b, 1941) to be of viral origin. This opinion has recently been confirmed (AIZAWA and FURUTA, 1964; AIZAWA, FURUTA, KURATA and SATO, 1964). The infectious agent, sensitive to the action of the silkworm's digestive fluid, is an icosahedral particle, 25—30 mμ in diameter. The symptoms of "clear heads" and diarrhea have been reproduced by infection with the virus and the disease can be diagnosed serologically by using an antiserum prepared with the isolated virus.

European Foulbrood of the Bee

One of the most discussed diseases, not only among bees, but also among insects in general, is European foulbrood. It has been distinguished from the beginning of this century from other diseases of the brood, and particularly from the American foulbrood, whose agent is *Bacillus larvae* (WHITE, 1907). The role of several bacteria such as *Bacillus pluton*, *Bacillus alvei*, *Streptococcus apis* and *Bacillus euridice*, was suspected, as were, recently, the interactions between such bacteria (WHITE, 1920; BAILEY, 1957, 1961; TOMASEC, 1961, 1965). Despite the therapeutic results obtained with chemicals or antibiotics, the mechanism of pathogenesis and etiology remained obscure and a viral etiology has been proposed (VAGO, 1964). Recently the presence of a paraspherical virus of 27—30 mμ was demonstrated in a highly infectious larval filtrate. The purified suspension of the virus was extremely infectious (GIAUFFRET, VAGO, ROUSSEAU and DUTHOIT, 1966). At present we do not know the exact role of this virus nor its relationship to the bacteria but it is probable that the natural disease is a complex syndrome resulting from viral and bacterial interaction.

Densonucleosis

One type of virus with a rather peculiar pathogenesis has been discovered in the lepidopteran *Galleria mellonella* L. (VAGO, MEYNADIER and DUTHOIT, 1964; MEYNADIER, VAGO, PLANTEVIN and ATGER, 1964). In Europe the industrial breeding stocks of this insect, used as a fish bait, were sometimes decimated by an extremely contagious disease. Paralysis and death followed 4 to 6 days after infection. The characteristic lesions, which are localized in the adipose tissue, consist of dense Feulgen positive masses inside the monstrously swollen nucleus. They are formed by the accumulation of icosahedral viruses (the smallest of invertebrate viruses) of 21 to 23 mμ, containing DNA. The particles are not regularly arranged and are mixed with a proteinaceous substance. Pathogenesis has also been obtained "in vitro" in cell cultures from the host and from *B. mori* (VAGO and LUCIANI, 1965; VAGO, QUIOT and LUCIANI, 1966). Inapparent virus infection was detected in larvae, nymphs, and adults by employing serum prepared with the purified virus (GIRAN, 1966). The virus appeared to be very specific for it has not been possible to infect any species other than the original host.

Virosis of Acarina

The first reported observations concerning mite virus diseases were due to viruses without inclusion bodies. Symptoms of paralysis have been observed in the red European mite *Panonychus ulmi* KOCH (STEINHAUS, 1959) and in the red citrus mite *Panonychus citri* MCGREGOR (SMITH, HILLS, MUNGER and GILMORE, 1959). In both cases, free viruses of 40 and 35 mμ, respectively, have been isolated. P. *citri* has been shown to contain DNA (ESTES and FAUST, 1965). While transmission trials with the first virus remained negative, infection has been obtained with the second by spraying the mites or by introducing diseased individuals into healthy populations (GILMORE, 1965). It is interesting to note that in the case of *P. citri*, numerous crystals accumulated in the hemocoele of sick mites as a result of metabolic disturbance (SMITH and CRESSMAN, 1962).

Iridescent Viroses

The observation of quite unaccustomed symptoms in the larvae of *Tipula paludosa* MEIGEN Diptera drew attention to a type of virosis, several varieties of which have been noted within a few years. The normal brown color is transformed to opalescent blue and this color appears in a particularly intense way on the lobes of adipose tissue (XEROS, 1954). A large number of viruses fills the cytoplasm of the adipose cells, those of the hypodermics and of the muscles. The relatively large icosahedral virus (130 mμ) (SMITH, 1956; SMITH and WIL-

LIAMS, 1958; WILLIAMS and SMITH, 1958) contains DNA (XEROS, 1954; SMITH, 1958; THOMAS, 1961; THOMAS and WILLIAMS, 1961). The purified viruses form a mass which itself becomes iridescent thanks to the regular arrangement of viruses in the tissue as in the pellet of ultra-centrifugation.

Another property of this virus is its low specificity. Its spectrum of virulence includes several species of Diptera, Lepidoptera and Coleoptera (BIRD, 1961; SMITH, HILLS and RIVERS, 1961; GERSHENSON, 1964).

A viral disease, similar in many ways to that of *Tipula* has been observed in *Sericesthis pruinosa* DALMAN (Coleoptera) (STEINHAUS and LEUTENEGGER, 1963). Adipose tissue and purified virus pelleted by centrifugation are iridescent and the isolated viruses are likewise icosahedral of 130 mμ, containing DNA, but serologically different from the virus of *Tipula* (DAY and MERCER, 1964). The iridescent effect and that of geometrical arrangement has been attributed to the presence of a surface layer of 50 mμ around the virus (MERCER and DAY, 1965). The development of this virus is similar to that of *Tipula* and it has been observed in the cytoplasm of cells of *Antheraea eucalypti* SCOTT cultured "in vitro" (BELLETT and MERCER, 1964; BELLETT, 1965 a, b).

The Lepidoptera have also furnished a case of iridescent virosis. A virus has been isolated from larvae of *Chilo suppressalis* WALKER showing a bluish tint. This virus was similar in all respects to the two viruses mentioned above, but it was slightly larger, 160 mμ (FUKAYA and NASU, 1966). The virus also developed in blood cells maintained "in vitro" (MITSUHASHI, 1966a, b.)

Finally, we shall mention the interesting discovery of iridescent virosis in mosquitoes, because, in this group of insects of medical importance, viral diseases may well have practical applications.

From iridescent larvae of *Aedes taeniorrhyncus* WIEDEMANN (CLARK, KELLEN and LUM, 1965) and of *Aedes fulvus pallens* WIEDEMANN, *Aedes vexans* MEIGEN and *Psorophora ferox* HUMBOLDT (CHAPMAN, CLARK, WOODARD and KELLEN, 1966), *Aedes anulipes* MEIGEN and *A. cantans* MEIGEN (WEISER, 1965) viruses of about 180 mμ have been isolated. The symptoms of disease resemble those described for the other iridescent viroses, but the structure and action of the viruses have not been studied in detail.

Lethargy

In the populations of the cockchafer *Melolontha melolontha* L. in the west of France, some larvae are delayed in development for several months. This state is transmissible by inoculation or by ingestion of affected tissue. From these, an icosahedral virus of 80 mμ approximately has been isolated, surrounded by a thick envelope and arranged in geometrical lines within the cytoplasmic clusters in the adipose tissue. This arrangement is reminiscent of that noted in the iridescent viruses, but no iridescent effect was noted either in the tissues or the mass of purified virus (HURPIN, VAGO, ROBERT and MEYNADIER, 1967; VAGO, DUTHOIT and ROBERT, 1967).

Flaccidity of *Antheraea eucalypti*

A disease with symptoms of flaccidity has been observed in larvae of the butterfly *Antheraea eucalypti* SCOTT (GRACE and MERCER, 1965). Dissection revealed that the mesointestinal tube was attacked, but spherical virus particles with a diameter of 50 mμ have been described from the haemolymph. Smaller particles were also found which might represent the liberated central cores. In histological sections the nuclei of mesodeum seemed to contain the virus.

Paralysis — Black Disease of the Bee

This concerns a group of symptoms which have long been observed in bees and attributed to various causes linked with their diet, hibernation, or intoxication. In certain cases, the insects fly with difficulty, their legs are paralyzed and tremble; in other they take on a dark tint while weakening. The two types have been considered either as independent or as linked with one another.

One histological characteristic has drawn attention to itself since the first advanced studies: the presence of granular inclusions in the cytoplasm of certain intestinal cells (MORISON, 1936). The infectious nature has been recognized (VECCHI and ZAMBONELLI, 1961) and a viral origin postulated. Later irregular oval elements approximately 27×45 mμ in diameter were isolated, and shown to be infectious (BAILEY, GIBBS and WOODS, 1963).

Similar viruses have been obtained from bees suffering from the "Forest diseases".

The presence of MORISON'S inclusions has also been shown in the nerve cells (GIAUFFRET, DUTHOIT and CAUCAT, 1966a).

In the cytoplasm of these cells corpuscles have been discovered, smaller than the purified virus (LEE and FURGALA, 1965c) or of the same size as the virus (GIAUFFRET, DUTHOIT and CAUCAT, 1966b).

Experimental Paralysis of the Bee

A viral action never encountered in natural conditions was noted after injection of extracts from healthy bees (BAILEY, GIBBS and WOODS, 1963). When such extracts, containing paraspherical particles of 28 mμ, were injected into healthy bees in doses greater than 13×10^2 particles, they produced acute paralysis in 2—4 days. Weak doses had no effect. Furthermore, the virus could be found regularly in small quantities in healthy bees (BAILEY, 1965). This "laboratory virosis" seemed to be localized in the adipose cells (FURGALA and LEE, 1966).

Aspects of Comparative Virology

In the preceding pages viruses of invertebrates causing a variety of diseases were described. It will be attempted to evaluate the characteristics of these viruses, first comparing them at the level of invertebrate virology, and later fitting them into the framework of general virology.

Pathogenesis

Non-occluded, free viruses are known to occur throughout the invertebrate world and they are infecting also groups in which inclusion-type viruses were not described. Thus, the first and only virosis noted among Orthoptera (MEYNADIER, 1966) the majority of viroses in Hymenoptera (BAILEY, GIBBS and WOODS, 1963; BŘČAK, 1964; GIAUFFRET, DUTHOIT and CAUCAT, 1966b; GIAUFFRET, VAGO, ROUSSEAU and DUTHOIT, 1966; VAGO, 1966a) and those of Coleoptera belong to the non-inclusion diseases (KRIEG and HUGER, 1960; STEINHAUS and LEUTENEGGER, 1963; HUGER, 1966). For invertebrates other than insects, the total is even more striking, for the only viroses known among Arachnidae (STEINHAUS, 1959; SMITH, HILLS, MUNGER and GILMORE, 1959), and Crustacea (VAGO, 1966b) are free virus diseases.

The tissue affinity is also markedly heterogenous. Similarly to free viruses, here also it is the adipose tissue which harbors the majority of viruses. The iridescent viruses, the densonucleosis virus and that of lethargy also belong here. Other viruses as the flaccidity virus seem to be selectively attached to the cells of the digestive tube. Finally, in the case of paralysis or black disease of the bee, lesions and even accumulations of viruses have been noted in the nervous tissue.

Both cytoplasmic and nuclear viruses have intracellular affinities. It is, however, interesting to note that among the free viruses discovered up to now, the great majority are localized in the cytoplasm. This group includes paralysis, the black disease and sacbrood in the bee, flaccidity, lethargy, all iridescent viruses, as well as all isolated paraspherical free viruses of crabs, crickets, and all Lepidoptera.

The only types of free viruses known to have an affinity for the nucleus are those of densonucleosis and, probably, the *Antheraea* flaccidity virus.

Development of viruses within the cell has been studied only in a few types of free viruses. Detailed studies have been made on iridescent viruses and densonucleosis.

In the group of iridescent viruses, studies have been carried out primarily on the extremely polyvalent virus of *Tipula paludosa*, inoculated into *Galleria mellonella* (Lepidoptera). Several divergent opinions have been advanced. According to some, preformed empty membranes may fill with a substance containing DNA, the latter growing outwards from the center (SMITH and HILLS, 1962). Another hypothesis concerns the role of a dense virogenic stroma which might produce fibrillas of nucleoprotein. At the same time open viral membranes might become free from microvesicles and fill up with nucleoprotein to form icosahedral bodies (XEROS, 1964).

BIRD's opinion (1962) contrasting in some aspects with the foregoing hypotheses, states that spherical bodies might shape from fibrous matter, and combine to form the dense central body of the virus. The envelope might form later from the residual fibrous matter.

An entirely different type of pathogenesis follows infection by densonucleosis. After the disappearance of all internal structures of the nucleus a homogenous area gives rise to paraspherical bodies which accumulate in a large mass, leaving a non-crystalline and probably non-solidified substance.

The hypothesis concerning the virus development of the transparent disease of Scarabeides is based on fragmentary facts only. The paraspheric virus particles seem to form at the expense of a virogenic stroma connected with abundant albuminoid masses in the cytoplasm of adipose cells (KRIEG and HUGER, 1960).

It was suggested that the virus of the Malaya disease of *Oryctes* could develop both in the nucleus and in the cytoplasm. The virus, whose elongated shape is considered as the mature form, could have originated equally from a virogenic stroma, from fibrillic or granular matter, or from immature spherical stages (HUGER, 1966).

Classification

The heterogenous character of the free viruses discussed poses the delicate question of relationship and of grouping. The general problem has been entrusted to a Sub-committee of the International Committee of Nomenclature of Viruses, specializing in invertebrate viruses. The task of this Sub-committee is all the more difficult because, until recently, invertebrate viruses have always been treated as a distinct group. This separation, without any doubt, arose from the rather special nature of polyhedra and of granules. Since the discovery of several free viruses, incorporation of insect viruses with other viruses seems easier.

We have to take into account the properties of viruses that have been accepted for the establishing of a general classification. These, although not yet officially defined, are related to intrinsic characteristics: nature of the nucleic acid, symmetry, envelope, number of capsomeres, diameter of the nucleocapsid. The resistance to hydrocarbons and to acids has also been envisaged. The mode of action and intracellular localization are considered as complementary notions.

If we examine (VAGO, 1966c), in the light of the demands made by the classification project, the data which we now possess on the character of the non occluded (free) viruses of invertebrates, we must admit that our knowledge is rather limited.

As the attached diagram shows the nature of the nucleic acid is often unknown. The symmetry is generally established, while the number of capsomeres is rarely defined. The presence of an envelope is often established. Other characteristics, such as the sedimentation constant, or resistance to chemical substances, are sometimes studied. On the other hand, affinity and intracellular localization are generally well known.

In order to classify free viruses of invertebrates, it is, above all, important and urgent to accumulate data about virus characteristics.

Aspects of Applied Virology

For a long time, the development of invertebrate virology was closely linked with the economic importance of certain diseases. Thus, early research

Table

Virus of	Nucl. Ac.	Strand.	Mol. W. of Nucl. Ac. in millions	% Nucl. Ac.	Particle Shape	Dim. Å.	Local. in Cell.	Host. (primary)
Iridescent of								
Tipula	DNA	2	140	15	Spher.	1,300	Cytopl.	Dipt.
Sericesthis	DNA	2	140	16	Spher.	1,300	Cytopl.	Coleopt.
Chilo	DNA	2	140		Spher.	1,600	Cytopl.	Lepidopt.
Mosquito					Spher.	1,800	Cytopl.	Dipt.
Flachery					Spher.	300	Cytopl.	Lepidopt.
Transparency	RNA				Spher.	600 to 750	Cytopl.	Coleopt.
Antheraea flaccidity	RNA				Spher.	500	Nuclear.	Lepidopt.
Densonucleosis	DNA	2	3		Spher.	230	Nuclear.	Lepidopt.
Bee chronic paralysis	RNA	1			Irreg.	270 × 450	Cytopl.	Hymeno.
Bee acute paralysis	RNA	1			Spher.	280	Cytopl.	Hymeno.
Sacbrood	RNA				Spher.	280 to 300	Cytopl.	Hymeno.
Cirphis					Spher.	250	Cytopl.	Lepidopt.
European foulbrood					Spher.	270		Hymeno.
Lethargy					Spher.	1,000	Cytopl.	Coleopt.
Citrus red mite	RNA				Spher.	350	Cytopl.	Acar.
European red mite					Spher.	400 to 600	Cytopl.	Acar.
Malaya disease					Elong.	2,000 × 700	Cytopl. or Nucl.	Coleopt.
Gryllus paralysis					Spher.	350		Orthopt.
Crab paralysis					Spher.	500 to 600	Cytopl.	Crustac.
Drosophila sigma								Dipt.

was almost exclusively concerned with the nuclear polyhedrosis of the silkworm and numerous studies are still being carried on even today on the various polyhedroses with a view to their applications in biological control.

Such reasons did not favor the study of free viruses. Recently, however, viruses of the non-occluded, free type have been linked to several important economic problems concerning both protection from harmful insects and the production of useful invertebrates.

Action on Harmful Insect Populations and Biological Control

Although our knowledge of the role of free viruses in the natural limitation of insect populations is more limited than that of viruses with inclusion bodies, several observations indicate the existence of *epizootics* and *enzootics*.

The iridescent virosis, found sporadically in England, has recently been shown to be an agent of persistent *enzootics* in certain *biotopes* of *Tipula paludosa* MEIGEN in the meadows of Normandy.

The virus of the "Malaya disease" of *Oryctes rhinoceros* L. seems to be a factor in the natural limitation of the population of this insect in the South Pacific. In fact, a pronounced death rate from this illness has been reported in Borneo, Fiji and Samoa (HUGER, 1966).

The transparent disease of Coleoptera *Melolontha melolontha* L. whose free virus etiology is established, has caused *epizootics* (HEIDENREICH, 1939; KRIEG and HUGER, 1960) in several biotopes in Germany.

From the human epidemiological point of view, particularly for the control of arthropod vectors, it would be interesting to obtain more information on viruses affecting mosquitoes. Present research in this field is only sporadic.

Free insect viruses do not seem suitable for biological control, because they lack the protection afforded by proteinaceous inclusion bodies. There were the polyhedra and granules, easy to purify, store and preserve without too many precautions, which gave a start to biological control with insect viruses. Of particular interest in this connection is the success obtained in the application of the densonucleosis virus against the invasion of beehives by the Lepidoptera *Galleria mellonella* L. (LAVIE, FRESNAYE and VAGO, 1965). It is true that, on the one hand, the virus in question is extremely contagious and fairly resistant despite the absence of inclusion bodies and, on the other hand, the beehive provides a special, closed, constant temperature environment.

Attempts have been made to find practical applications for the free virus of the red acarian of the citrus-tree, *Panonychus citri* (MCGREGOR). Sprays of fresh suspensions or the introduction of contaminated acarians into the natural populations of that insect have lowered population levels (GILMORE, 1965).

It would be useful to examine the prejudice against the possibility of using free viruses in biological control by revising present, classic conceptions, on the mechanism of action in biological control. In fact, recent experiments

(VAGO, FOSSET and BERGOIN, 1966; BERGOIN, 1966) have shown the role which both cannibalism and the passage through predator and entomophagous insects can play in the rapid spread of viruses, including those without inclusion bodies, for instance, densonucleosis virus.

Hygiene of Useful Invertebrates

The discoveries made in the last few dozen years have shown the particularly important role played by free viruses in the rearing of useful animals, or at least in that of two classic "domesticated" insects: the silkworm and the honey bee.

As for the first, the recognition in the last 40 years (PAILLOT, 1930a, b, 1941; AIZAWA and FURUTA, 1964; AIZAWA, FURUTA, KURATA and SATO, 1964) of the virus etiology of one type of flaccidity constituted an advance in the campaign against the "flaccidity" group of diseases. The nature of these diseases has been controversial and economically important in sericulture.

Regarding the bee, at least six diseases which have been observed for a very long time by bee-keepers, have been determined etiologically by the isolation of free viruses: sacbrood, paralysis, the black disease, atypical weakening, and, probably, European foulbrood (WHITE, 1917; VECCHI and ZAMBONELLI, 1961; BAILEY, GIBBS and WOODS, 1963, 1964; VAGO, 1964, 1966a; GIAUFFRET, DUTHOIT, CAUCAT, 1966b; GIAUFFRET, VAGO, ROUSSEAU and DUTHOIT, 1966). Excepting European foulbrood, all these illnesses are considered minor since bees often recover and harmful effects do not noticably diminish the bees' production. All the same, they occur frequently and often contribute to the deterioration of hives. In European foulbrood, anti-bacterial therapy is practised because of complications which render this illness really grave. Antiviral prophylaxy for the various diseases has not been worked out.

Conclusions

The various aspects described show that research on free insect viruses of invertebrates play an increasingly important role in invertebrate pathology. Thanks to them, our virological knowledge is spreading to classes of invertebrates in which until recently no virus has been observed, and to types of virus never before detected in invertebrates.

Apart from invertebrate pathology, we wish to emphasize the role played by free viruses in the narrowing of the gap between insect virology and general virology. Invertebrate viruses have long been considered as a separate group, not only because of the characteristics of their hosts, but above all because the viruses with the polyhedral or granular type of inclusion bodies have no equivalent among the viruses of vertebrates or those of plants. The formation of proteinaceous bodies with a crystal-like structure in which the viral bodies insert themselves progressively constitutes a pathogenesis characteristic only for insect viruses.

The discovery of paraspherical viruses, found free or in clusters in the cytoplasm or the nuclei, revealed the existence among insects of viruses similar to those of vertebrates or plants. Comparisons of their structures and properties have since been facilitated.

Relationships with general virology are also being established by the discovery of viruses with particular structures and pathogenesis. Thus, the viruses of iridescent disease, of the lethargy, or of the densonucleosis have widened the spectrum of viruses by introducing new types.

Finally, from the comparison of data furnished by general virology and by that of invertebrates, we note the difficulty of distinguishing between categories of free viruses and those with inclusions. In effect, over and beyond the types freely scattered in the cytoplasm and the nucleus (like the paralyses and flaccidity), we observe diseases like densonucleosis in which the viruses agglomerate in the nucleus in a compact mass which also includes a linking substance; in lethargy, or the iridescent viroses, the viruses collect in clusters.

Thus, in our opinion, even in putting forward the definition of "inclusion body" and of "clusters of viruses" at the level of comparative virology we must remain aware of the fact that, in reality, infinite intermediate states of definition and form are to be expected between the free viruses and the different types of inclusion-type viruses.

References

AIZAWA, K., and Y. FURUTA: C. R. Coll. Int. Pathol. Insectes, Paris; 1962. Entomophaga mém. H.S. n° **2**, 373—376 (1964).
— — K. KURATA, and F. SATO: Bull. Sericult. Exp. Sta. **19**, 223—240 (1964).
BAILEY, L.: Nature (Lond.) **180**, 1214—1215 (1957).
— Bull. Apicole **4**, 111—112 (1961).
— J. Invert. Pathol. **7**, 167—169 (1965).
—, A. J. GIBBS, and R. D. WOODS: Virology **21**, 390—395 (1963).
— — — Virology **23**, 425—429 (1964).
BELLETT, A. J. D:. Virology **26**, 127—31 (1965a).
— Virology **26**, 132—41 (1965b).
—, and E. H. MERCER: Virology **24**, 645—653 (1964).
BERGOIN, M.: Thèse Faculté des Sciences de Marseille (1966).
BIRD, F. T.: Canad. J. Microbiol. **7**, 827—830 (1961).
— Canad. J. Microbiol. **8**, 533—534 (1962).
BŘČAK, J.: Bull. Apicole **7**, 119—121 (1964).
—, and O. KRÁLÍK: J. Invert. Pathol. **7**, 110—111 (1965).
—, J. SVOBODA, and O. KRÁLÍK: J. Insect Pathol. **5**, 385—86 (1963).
CHAPMAN, H. C., I. B. CLARK, B. WOODWARD, and W. R. KELLEN: J. Invert. Pathol. **8**, 545—546 (1966).
CLARK, T. B., W. R. KELLEN, and P. T. M. LUM: J. Invert. Pathol. **7**, 519—521 (1965).
DAY, M. F., and E. H. MERCER: Aust. J. biol. Sci. **17**, 892—902 (1964).
ESTES, Z. E., and R. M. FAUST: J. Invert Pathol. **7**, 259 (1965).
FUKAYA, M., and S. NASU: Appl. Entomol. Zool. **1**, 69—72 (1966).
FURGALA, B., and P. E. LEE: Virology **29**, 346—348 (1966).
GERSHENSON, S.: C.R. Coll. Int. Pathol. Insectes. Paris 1962. Entomophaga mém. H.S. n° **2**, 361—372 (1964)

GIAUFFRET, A., J. L. DUTHOIT, and M. J. CAUCAT: Bull. Apicole 9, 221—228 (1966a).
— — — Rec. Méd. vét. 142, 819—829 (1966b).
— C. VAGO, M. ROUSSEAU, and J. L. DUTHOIT: Bull. Apicole 9, 123—134 (1966).
GILMORE, J. E.: J. econ. Entomol. 58, 1136—40 (1965).
GIRAN, F.: Thèse présentée à la Faculté de Sciences de l'Université de Montpellier (1966).
GRACE, T. D. C., and E. H. MERCER: J. Invert. Pathol. 7, 241—244 (1965).
HEIDENREICH, E.: Proc. 7th Int. Congr. Entomol. Berlin, vol. 3, p. 1905—2171 (1939).
HOLMES, F. O.: In: Bergey manual of determinative bacteriology, 6th ed., p. 1225—1228. Baltimore: Williams & Wilkins Co. 1948.
HUGER, A. M.: J. Invert. Pathol. 8, 38—51 (1966).
HURPIN, B., C. VAGO, P. ROBERT, and G. MEYNADIER: Entomophaga 12, 311—315 (1967).
KRIEG, A., and A. HUGER: J. Insect Pathol. 2, 274—288 (1960).
LAVIE, P., J. FRESNAY, and C. VAGO: Ann. Abeille 8, 321—323 (1965).
LEE, P., and B. FURGALA: Virology 25, 387—392 (1965a).
— — J. Invert. Pathol. 7, 502—505 (1965b).
— — J. Invert. Pathol. 7, 170—174 (1965c).
MERCER, E. H., and M. F. DAY: Biochim biophys. Acta (Amst.) 102, 590—599 (1965).
MEYNADIER, G.: C.R. Acad. Sci. (Paris) 266, 742—744 (1966).
—, C. VAGO, G. PLANTEVIN, and P. ATGER: Rev. Zool. agricole et appl. 63, 207—208 (1964).
MITSUHASHI, J.: Appl. Entomol. Zool. 1, 130—137 (1966a).
— Appl. Entomol. Zool. 1, 199—201 (1966b).
MORGENTHALER, O.: Schweitz. Bienenztg. 70, 148—151 (1947).
MORISON, G. D.: Rothamsted conferences 12, 17—21 (1936).
PAILLOT, A.: C.R. Soc. Biol. (Paris) 104, 585—586 (1930a).
— Traité des maladies du ver à soie, 279 p. Paris: Gaston Doin 1930b.
— Ann. Epiphyt. 7, 99—117 (1941).
PASTEUR, L.: C. R. Acad. Sci. (Paris) 66, 1289 (1868).
SMITH, K. M.: J. biophys. biochem. Cytol. 2, 301—302 (1956).
— Parasitology 48, 459—462 (1958).
—, and A. W. CRESSMAN: J. Insect Pathol. 4, 229—236 (1962).
—, and G. J. HILLS: C.R. XI Int. Kongr. Entomol. Wien, 1960, p. 823—827. 1962.
— — F. MUNGER, and J. E. GILMORE: Nature (Lond.) 184, 70 (1959).
— —, and C. F. RIVERS: Virology 13, 233—241 (1961).
—, and R. C. WILLIAMS: Endeavour 17, 12—21 (1958).
STEINHAUS, E. A.: Bact. Rev. 13, 203—223 (1949).
— Hilgardia (Berkeley, Calif.) 20, 629—678 (1951).
— J. Insect Pathol. 1, 435—437 (1959).
—, and R. LEUTENEGGER: J. Insect Pathol. 5, 266—270 (1963).
THOMAS, R. S.: Virology 14, 240—252 (1961).
—, and R. C. WILLIAMS: J. biophys. biochem. Cytol. 11, 15—29 (1961).
TOMASEC, I.: Bull. Apicole 4, 2 (1961).
— Veterinarski Archiv (Zagreb) 35, 11—20 (1965).
VAGO, C.: Bull. Apicole 7, 111—118 (1964).
— Bull. Apicole 9, 5—9 (1966a).
— Nature (Lond.) 209, 1290 (1966b).
— Entomophaga 11, 347 (1966c).
— H. DUTHOIT, and P. ROBERT: C.R. Acad. Sci. (Paris) (in press).
— J. FOSSET, and M. BERGOIN: Entomophaga 11, 177—182 (1966).
—, and J. LUCIANI: Experientia (Basel) 21, 393 (1965).
— G. MEYNADIER, and J. L. DUTHOIT: Ann. Epiph. 5, 475 (1964).

VAGO, C., J. M. QUIOT, and J. LUCIANI: C.R. Acad. Sci. (Paris) **263**, 799—800 (1966).
VECCHI, M. A., and C. ZAMBONELLI: Bull. Apicole **4**, 181—198 (1961).
WASSER, H. B.: J. Bact. **61**, 787—792 (1952).
WEISER, J.: Bull. Org. mond. Santé **33**, 586—588 (1965).
WHITE, G. F.: Bull. U.S. Dept. Agric. Bur. Ent. Circ. **94**, 4 p. (1907).
— U.S. Dept. Agr. Bur. Ent. Circ. **169**, 5 p. (1913).
— Bull. U.S. Dep. Agric. exp. Stat. **431**, 54p. (1917).
— Bull. U.S. Dep. Agric. exp. Stat. **810**, 46p. (1920).
WILLIAMS, R. C., and K. M. SMITH: Biochim. biophys. Acta (Amst.) **28**, 464—469 (1958).
XEROS, N.: Nature (Lond.) **174**, 562 (1954).
— J. Insect Pathol. **6**, 261—283 (1964).

Arboviruses, the Arthropod-Borne Animal Viruses *

Roy W. Chamberlain **

Arboviruses are unique in infecting both vertebrates and blood-sucking arthropods. Acquired from the infected blood of vertebrates, they multiply in the arthropod, become established in the salivary glands, and are subsequently transmitted to other susceptible vertebrates by bite. In briefest terms, arboviruses are defined as animal viruses transmitted biologically by hematophagous arthropods to vertebrate hosts (Casals, 1966; Andrewes, 1962).

To date, 204 arboviruses have been catalogued, which makes them by far the largest group of vetrebrate viruses known (Cabv, 1967). They are worldwide in distribution, with at least some types found wherever a search has been made and suitable vectors and hosts abound, and include among their numbers the causative agents of some of the most important epidemic diseases of man.

Properties and Classification

General Properties of Arboviruses

The definition of arboviruses does not necessarily restrict them to a homogeneous group, either by similarity of structure or line of origin. It is possible that they could contain either RNA or DNA. They could be small or large, spherical or rod-shaped, enveloped or naked. It is to be expected, therefore, that they should make up a hodgepodge group. Surprisingly, a degree of order higher than expected appears to exist. RNA has been found in all arboviruses in which determinations have been made, about 20 (Casals, 1966), and studies on a similar limited number show the majority to be spherical. Most of them range in size from 20—100 mμ, although a few are considerably larger (150 mμ or more). One nontypical group is rod-shaped, however, measuring 180—210 mμ by 50—70 mμ, with helical symmetry (Howatson and Whitmore, 1962; Mussgay and Weibel, 1963). Steady progress is being made in this area, but a much greater knowledge of virion structure of the individual members will be required before an accurate classification based upon a universal system, such as proposed by Lwoff et al. (1962), will be possible.

The arboviruses in general are heat and acid labile and are stabilized by serum proteins. A large number have been tested for sensitivity to ether or

* U.S. Department of Health, Education and Welfare Public Health Service. Bureau of Disease Prevention and Environmental Control, National Communicable Disease Center, Atlanta, Georgia 30333.

** Chief, Arbovirus Infections Unit, Laboratory Improvement Program.

sodium deoxycholate and the majority shown to be inactivated, indicating presence of a lipid-containing coat or envelope associated with infectivity (CABV, 1967; ANDREWES and HORSTMANN, 1949; THEILER, 1957; FRANKLIN, 1962). Essentially all are pathogenic for suckling white Swiss mice by the intracerebral route, although this is to a large extent a selectively derived property since i.c. inoculation of suckling mice has been the method generally used in recent years for isolation of new arbovirus strains. It is quite likely that many arboviruses exist which cannot be detected by suckling mouse inoculation, but in the absence of a more sensitive detection method, these have not been disclosed.

Classification Based Upon Antigenic Relationship

About three-fourths of the arboviruses currently recognized have been placed into 22—26 antigenic groups, based upon cross-reactions in one or more of three basic serologic tests — hemagglutination-inhibition (HI), complement fixation (CF) and neutralization (NT). In general the HI test shows the broadest crossing, the CF, an intermediate amount, and the NT, the least. For detecting antigenic relationships among different arboviruses, broad crossing is highly advantageous; however, for specific virus identification and separation from allied forms, the tests with narrower limits are desired (CASALS and REEVES, 1965; CASALS and BROWN 1954; CASALS, 1957; YARU, 1966).

According to the system, each member of an arbovirus group should indicate relationship to other members of the group by serological crossing, but should not cross to a significant extent with members of other groups. At least two viruses are required to make up a group. Those not crossing with any other known arboviruses are listed as "Ungrouped".

The first three groups were given the letter designations A, B, and C. Subsequent groups were named from their type viruses, e.g., Bunyamwera group, Guama group. It has become customary to name arboviruses after the geographic areas of their discovery (Semliki Forest, Cache Valley), although some names stem from disease symptoms produced (yellow fever, louping ill), and others are a combination of locality and disease symptom (West Nile and St. Louis encephalitis). A few derive their names from other sources, such as the arthropod or vertebrate yielding the first isolation, giving us such examples as Trivittatus and U.S. bat salivary gland viruses.

A provisional listing of presently recognized arbovirus groups, with representative viruses of each, is given in Table 1. For more complete details on the individual viruses and their properties one should consult the Arbovirus Catalogue sponsored by the American Committee of Arthropod-borne Viruses (1967), the Rockefeller Foundation Virus Laboratory annual reports and several other basic sources of reference (CASALS and CLARKE, 1965 a; CLARKE and CASALS, 1965; CASALS and CLARKE, 1965 b).

Group B is the largest with about 40 member viruses. Some of its members show especially close serological relationship to each other, and thus form subgroups or complexes within the large group. The Japanese B — West Nile

Table 1. *Provisional antigenic grouping of arboviruses, 1967*[a]

Group	Examples	Approximate No. in group	Some of group known to cause human disease
A	Eastern, Western, Venezuelan encephalitis	20	+
B	St. Louis, Japanese B, West Nile encephalitis, yellow fever, Russian spring-summer encephalitis	40	+
Bunyamwera supergroup			
Bunyamwera	Bunyamwera, Guaroa, Cache Valley	13	+
Bwamba	Bwamba, Pongola	2	+
C	Apeu, Caraparu, Itaqui, Madrid	10	+
California	California, Trivittatus, Tahyna	6	+
Capim	Capim, Acara, Bushbush	5	—
Guama	Guama, Catu, Bimiti, Moju	5	+
Simbu	Simbu, Buttonwillow, Oropouche	10	+
Unassigned	Mirim, Patois, Sontecomapan, Zegla	4	—
Anopheles A	Anopheles A, Lukuni	2	—
Anopheles B	Anopheles B, Boracea	2	—
Bakau	Bakau, Ketapang	2	—
Changuinola	Changuinola, Irituia	3	+
Epizootic hem. dis. of deer	EHD-N.J., EHD-S.D.	2	—
Hughes	Hughes, Farrallon	2	—
Kaisodi	Kaisodi, Lanjan	2	—
Kemerovo	Kemerovo, Chenuda, Tribec	3	+
Koongol	Koongol, Wongal	2	—
Nyando	Nyando	2	—
Phlebotomus Fever	Naples sandfly fever, Sicilian SF, Chagres, Candiru	10	+
Piry	Piry, Chandipura	2	+
Quaranfil	Quaranfil, Johnston	2	+
Tacaribe	Tacaribe, Amapari, Junin, Machupo	4	+
Timbo	Timbo, Chaco	2	—
Turlock	Turlock, Umbre	2	—
Vesicular Stomatitis	VSV-N.J., VSV-Ind., Cocal	4	+
Ungrouped	Rift Valley fever, Colorado tick fever, Nyamanini, Crimean hemorrhagic fever, Pacui, Lagos bat	50	+

[a] Assembled largely from data in Yale Arbovirus Research Unit Annual Report, 1966. The listing is admittedly incomplete in view of the rapid developments in this field, but serves to illustrate the variety of viruses represented.

complex is an important one, which also includes the viruses of St. Louis and Murray Valley encephalitis and Ilheus. At least four dengue viruses form their own complex, and the tick-borne viruses, still another.

Complexes are less evident in Group A, which seems to lack a common antigenic base found in the Group B; the individual members thus appear more distinct from each other. About 20 members are in this group.

The next largest group is Bunyamwera, which along with a number of other groups (C, Bwamba, California, Capim, Guama, Simbu and Unassigned) make up the Bunyamwera supergroup comprising more than 50 different viruses (YARU, 1966). Numerous other smaller groups or ungrouped viruses comprise the remainder.

Criteria for Defining Arboviruses other than by Proving Biological Transmission

In the early days of arbovirus research, the few virus isolations made were from naturally occurring cases of disease in man or animals, generally from autopsy tissues or blood specimens. The virological methods employed were not particularly sensitive, which limited the kinds of viruses isolated. Prior to 1940, only 15 different arboviruses had been recognized (CABV, 1967). In view of their restricted number and obvious medical or veterinary importance, most of them received thorough study and were proven to be arthropod-borne by vector transmission tests.

More recently, however, with application of improved isolation methods to broad field surveys in many parts of the world, the new number of arboviruses discovered has increased at a phenomenal rate. It has jumped from 15 in 1940 to 34 in 1950, 123 in 1960 and 204 in 1967 (CABV, 1967). Although human and animal diseases have continued to contribute to this score, the greater share of new isolations has been from field-collected, blood-sucking arthropods, the blood or tissues of wild animals, and from susceptible sentinel animals, usually suckling mice or hamsters, exposed in the field to the bites of various kinds of arthropod vectors.

This great flood of new viruses has made it virtually impossible to conduct vector transmission studies on them all to determine if they truly qualify as arboviruses; in fact, complete studies have been carried out on only a relatively small proportion. As a practical solution to this problem, various other criteria have been used (ANDREWES, 1962), which in most instances have made direct proof of biological transmission appear unnecessary. The more applicable criteria are as follows:

1. Antigenic relationship to a previously known arbovirus. With the present classification system of grouping based upon serologic crossing, this is accepted as proof.

2. Isolation from field-collected, blood-sucking arthropods. This alone does not prove a virus to be arthropod borne, but provides a very strong indication, particularly if repeated isolations are made from a species or group or in association with infected animals.

3. Isolation from sentinel animals exposed in the field to the bites of blood-sucking arthropods. This evidence is even stronger than isolation from an arthropod alone, provided the sentinels were exposed in such a manner that disease could logically be acquired only from the bites of arthropod vectors.

4. Infectivity for mosquitoes or other vector groups by feeding or intrathoracic inoculation, with infection of the salivary glands. The salivary glands are generally dissected out after a reasonable incubation period and tested for virus. Its presence indicates probable ability of the species to transmit by bite.

5. Isolation from cases of disease in man or animals occurring under circumstances epidemiologically suggesting arthropod transmission.

6. Inactivation by exposure to ether or sodium deoxycholate, since the greatest share of known arboviruses possess this characteristic.

7. Size compatible with that of most known arboviruses.

8. Presence of RNA.

9. Heat and acid sensitivity.

10. Pathogenicity for suckling mice inoculated by the intracerebral route.

Some of these criteria alone are inadequate, but in combination with others provide reasonable bases for classifying new virus isolants as arboviruses.

The Arthropod Vectors

The Types of Arthropod Vectors and Vector Specificity

Only four groups of arthropods are known to play important vector roles: mosquitoes, ticks, sandflies (*Phlebotomus*) and certain blood-sucking gnats (*Culicoides*). Lelaptid mites are also suspect for some viruses, but final proof is still lacking. Mosquitoes are by far the most important and probably serve as vectors for at least two-thirds of all of the known arboviruses. The ticks, sandflies and gnats transmit most of the remainder, with the ticks currently claiming the greater share. A more definite statement cannot be made since vector studies have been completed on only a fraction of the total number of arboviruses; the type of vector for a large proportion is assumed on the basis of relationship to more completely studied viruses or by circumstances of isolation.

The effective vector spectrum in nature for a particular arbovirus is generally narrow; within a geographic area it may be restricted to a single species or complex. Part of this specificity is probably due to greater susceptibility and transmitting efficiency of the primary vector species than of other potential vectors in the area, and part to especially favorable ecological relationships with the essential vertebrate hosts.

Mosquitoes are the main vectors for all Group A, most of Group B, the Bunyamwera supergroup, some of the minor groups, and about a dozen viruses at present ungrouped. The ticks are vectors for seven or eight viruses in the tick-borne complex of Group B, nine viruses in four recently erected small groups (Hughes, Kaisodi, Kemerovo and Quaranfil), and probably about 18 ungrouped viruses, the best known member of which is Colorado tick fever virus.

Sandflies (*Phlebotomus*) are known vectors for the Phlebotomus group of viruses, comprising about ten members. The members of this group are probably more abundant than the relatively few isolations have indicated. Comparatively little effort has been devoted to *Phlebotomus* collecting and testing; furthermore, some of the virus strains have shown low pathogenicity for suckling mice, with serial blind passaging sometimes necessary to reveal their presence. *Phlebotomus spp.* have also yielded isolations of vesicular stomatitis virus (SHELOKOV et al., 1961).

Definitive arbovirus studies with *Culicoides* gnats have thus far been restricted to the blue tongue complex of sheep and African horse sickness. However, eastern encephalitis virus (Group A) has been isolated from them in nature on one occasion (KARSTAD et al., 1957), and recent findings in California suggest that they may be vectors for two other viruses (REEVES, 1967).

The role of blood-sucking mites as arbovirus vectors is at present unresolved. The viruses of western encephalitis (Group A) and St. Louis encephalitis (Group B) have been isolated from mites collected in avian nests and roosts, but evidence of their ability to transmit has been unconvincing (SMITH et al., 1944; REEVES et al., 1955; SULKIN et al., 1955; CHAMBERLAIN and SIKES, 1955; CHAMBERLAIN et al., 1957). Rodent mites have yielded Cocal virus (vesicular stomatitis group) in Trinidad (JONKERS et al., 1964) and Junin virus (Tacaribe group) on a number of occasions in Argentina (CASALS and CLARKE, 1965 b). In no instance, however, has proof of ability to transmit effectively been demonstrated. Best judgment suggests that mites are unimportant as vectors of the viruses in Groups A and B. The status of mites as vectors for the vesicular stomatitis and Tacaribe groups certainly requires more study. Flanders virus, apparently a member of the vesicular stomatitis group (MURPHY et al., 1966), is assuredly mosquito-borne, as evidenced by numerous isolations from bird-feeding mosquitoes (CHAMBERLAIN, 1966). Also, recent experimental studies have shown vesicular stomatitis virus to multiply in and be transmissible by *Aedes aegypti* mosquitoes (MUSSGAY and SUAREZ, 1962; BERGOLD, 1967). There is a good possibility that the Tacaribe group requires no arthropod vectors at all, since Machupo virus has been shown to be excreted in the urine of the infected wild rodent host (JOHNSON et al., 1965).

Virus Behavior in the Arthropod Vector

Knowledge of the behavior of arboviruses in their invertebrate hosts has come mostly from studies conducted with mosquitoes, but there is reason to believe the findings generally hold true for the other vector groups as well. The ticks appear to exhibit some significant differences, however. In them, virus development may be slower, transovarial and transstadial transmission may occur, and conceivably virus in the coxal fluid could be an additional means of transmitting infection to vertebrates through skin contamination. Mites, if proven to be significant vectors, possibly would follow some of the patterns set by the ticks.

The successful completion of a cycle of infection is dependent upon an adequate level of virus in the vertebrate host blood, ingestion of the virus by a susceptible vector species, gut infection of the vector, distribution of virus via the hemolymph to various parts of the body including the salivary glands, growth in the glands, and finally injection into a susceptible vertebrate host during a subsequent blood feeding. Viremias vary in intensity in different vertebrate host species and may be present for brief or long periods. The susceptibility of different arthropod vectors varies also. Much is dependent upon close association of vector and natural host to make initial infection possible and to allow opportunity for the refeedings necessary for continued virus propagation. A high replacement rate of the host is advantageous by assuring a replenishing supply of susceptibles since antibody build-up undoubtedly serves to limit the virus cycle. An adequate supply of the vector species is also required, and this is dependent upon favorable climatic and breeding conditions.

In mosquitoes, at least, a threshold phenomenon is evident in infection of the gut (CHAMBERLAIN et al., 1954). A critical concentration of virus in the blood meal is required to establish gut infection in a significant proportion of the vectors. Below this concentration, infection is rarely established; above it, it is established with proportionally greater frequency. The threshold varies for different viruses and different vector species. The difference between vector species is probably due to variation in the number of suitable receptor sites on the cells lining the gut. A highly susceptible vector has a low threshold of infection; a less susceptible one, a higher treshold. It is significant that once the threshold "barrier" is overcome and infection established, the virus develops with about equal ease in either low- or high-threshold species. Usually a longer extrinsic incubation period is required of weak, bare-threshold infections, however, since more cycles of replication are required for virus build-up.

In the field a low-threshold vector generally should have a better chance to become infected than one with a high threshold and should be more apt to play a significant role as an epidemic vector, since a greater variety of hosts may serve to infect it. A high-threshold vector could assume importance, however, if closely associated with high-viremia hosts or if present in large numbers.

When virus has progressed through the gut cells, it is distributed by the arthropod hemolymph to various parts of the body. Whether or not virus multiplies in cells of the hemolymph is unknown; the concentration of virus in the gut tissue is high, and could easily account for the levels found in the hemolymph. Infection throughout the body is established in various dissimilar organs and tissues: ovaries, nerve tissue and, if the vector is to be able to transmit by bite, in the salivary glands (LAMOTTE, 1960). There is little doubt that when infection is established in the salivary glands, virus actually multiplies in the cells of the glands (CHAMBERLAIN and SUDIA, 1961). An infection threshold phenomenon has not been as clearly defined for the salivary glands as for the gut. Absence of virus in the hemolymph, through failure to pass through the gut wall, could account for poor salivary gland infections.

"Typical" arboviruses are generally believed to produce no cellular damage in their appropriate vector species. Radical salivary gland changes have been noted in *Culicoides* infected with blue tongue virus (BOWNE and JONES, 1966), but there is some doubt whether blue tongue is a typical arbovirus in view of its resistance to inactivation with ether or sodium deoxycholate (SVEHAG et al., 1966). Semliki Forest virus has recently been shown to cause lesions in the salivary glands of *Aedes aegypti*, however (MIMS et al., 1966). Although this species is not the true vector of Semliki Forest virus, this finding indicates the need for a review of the status of cytopathogenicity of arboviruses for vector groups. Damage to salivary gland cells could greatly reduce the transmitting efficiency of a species and make it a poor vector.

The extrinsic incubation period (time required for virus to complete development in the vector) varies with the kind of virus and the species of vector. Some viruses grow more quickly than others. For example, eastern encephalitis virus has been shown to increase about ten fold each eight hours in an efficient vector species (CHAMBERLAIN and SUDIA, 1957). St. Louis encephalitis virus, on the other hand, also in an efficient vector species, required about 12 hours for each ten-fold replication (Sudia, 1959). Some vector species apparently provide a better medium for rapid replication than others. *Culex pipiens pipiens* was able to transmit St. Louis encephalitis virus with peak efficiency several days earlier than *C. p. quinquefasciatus* infected and incubated concurrently under the same conditions (CHAMBERLAIN et al., 1959; CHAMBERLAIN et al., 1966). As another example, one strain of *Aedes aegypti* infected with eastern encephalitis virus required 13 days of incubation at 80° F to attain a transmission rate of 56 percent (CHAMBERLAIN et al., 1954), whereas a different strain of the same mosquito species was able to transmit the identical virus with a rate of 89 percent after only six days of incubation (CHAMBERLAIN, 1966). Possibly in the "faster" mosquito species, fewer cycles of virus replication were required for passage through the gut wall, or salivary gland infection was established more easily. In the case of the extremely short extrinsic incubation period in the *A. aegypti* strain, direct diffusion of virus through gut cell membranes is suggested (BOORMAN, 1960).

The rate of virus development is in direct proportion to the temperature of incubation (DAVIS, 1932; CHAMBERLAIN and SUDIA, 1955). The extrinsic incubation period may range from only a few days at summer temperatures of over 90° F to several weeks at temperatures in the 60's and low 70's.

Infection is generally considered to last for the life of the vector, although ability to transmit may decrease in old age. In mosquitoes, virus gradually dies out in the organs first infected, such as the gut and ovaries, but apparently leaves the salivary glands last of all (LAMOTTE, 1960) unless, as has recently been indicated with Semliki Forest virus in *Aedes aegypti*, the cells of the glands are destroyed by the infection (MIMS et al., 1966).

Arthropods as Arbovirus Reservoirs

The arthropods, with the exception of ticks, have shortcomings as overwintering reservoirs. No proof of effective transovarial passage of virus has yet been obtained for mosquitoes, and overwintering infections in adult hibernating mosquitoes appear unlikely to be efficient enough to account for perpetuation of viruses. Also, the mosquito vectors of many arboviruses do not overwinter as adults.

Transovarial transmission of viruses by *Phlebotomus* is indicated but not firmly substantiated (CASALS and CLARKE, 1965b). At best, this would explain the reservoir mechanism for only one virus group. Even with the ticks, the effectiveness of transovarial transmission may be overrated. It has not been proven for Colorado tick fever virus and from the few reports concerning other tick-borne viruses, this manner of transmission does not appear to be highly efficient (HOOGSTRAAL, 1966). It is unquestioned, however, that transstadial transmission and life-long infection in ticks occur, and probably serve most effectively as reservoir mechanisms for the viruses involved.

The Vertebrate Hosts

The Types of Vertebrate Hosts

With few exceptions, the natural vertebrate hosts of the arboviruses are wild animals. There seem to be arboviruses for almost any kind of animal, including rodents, birds and reptiles. The host spectra for some viruses, such as those of eastern and western encephalitis, are very broad, including a great many species and families, and may even cross class lines. The spectrum of some others may be quite narrow, however; the effective hosts of yellow fever, for example, are apparently limited to man and monkeys, and of dengue, perhaps man alone.

Birds are particularly involved with a large number of the mosquito-borne encephalitogenic viruses, for example, eastern and western encephalitis viruses of Group A and several of Group B (St. Louis, Japanese B, Murray Valley and West Nile). Also they may play a significant role in the dissemination of tick-borne viruses, even when the main hosts are rodents, both by transporting the ticks and by serving as viremic hosts in some of the transmission cycles. Ticks of many species have been collected from migrating birds on important fly-ways of the world (HOOGSTRAAL, 1966). Birds are apparently not suitable hosts for many other viruses, however, such as many members of the Bunyamwera supergroup which appear primarily to infect rodents.

The host spectra of a large proportion of the arboviruses are still little understood, mainly because the detailed studies required to obtain this knowledge have not been conducted. Antibody studies on field-collected sera have been done in many instances, but have often failed to distinguish between important and unimportant hosts since many incidentally infected and insignificant species may be included. Accurate assessment of a vertebrate's role as an arbovirus host in nature is dependent both upon detailed ecologic observations and comprehensive laboratory studies.

Some arboviruses are apparently restricted to cold-blooded vertebrates; three (Chaco, Marco and Timbo) have been isolated only from tropical lizards

(CAUSEY et al., 1966). It is possible to infect snakes and turtles with some viruses commonly thought to be restricted to warm-blooded hosts; experimental infections with eastern and western encephalitis viruses have been studied (KARSTAD, 1961; HAYES et al., 1964; THOMAS et al., 1958; THOMAS and EKLUND, 1960; THOMAS and EKLUND, 1962). Isolations of the latter virus from wild-caught garter snakes have been reported in Utah and Canada (GEBHARDT et al., 1964; BURTON et al., 1966).

The Course of Infection in Vertebrates

Peripheral inoculation of virus into a susceptible warm-blooded host apparently initiates infection in the endothelial cells of blood vessels and in the blood-forming organs, such as bone marrow, and in some instances, the liver. A viremia occurs, and generally attains high levels within the first 24—72 hours, then gradually decreases. Depending upon the type of virus and the kind of animal involved, the viremia may be of brief duration (up to five days) or may extend over longer periods. With the viruses infectious for birds the viremias generally last no more than a week, and are in levels capable of infecting preferred vectors for only a portion of that time. Antibodies commonly appear promptly at cessation of viremia. Infections produced in rodents by their various arboviruses are generally assumed to fall into the same pattern as in birds since the species tested have usually also exhibited limited viremias and prompt antibody responses. However, the dormouse and hedgehog infected with central European tick-borne encephalitis virus have been reported capable of carrying a viremia through winter hibernation if onset of cold weather is at a critical stage of virus development (HOOGSTRAAL, 1966). Extended viremias of Tahyna virus in hibernating hedgehogs have occurred as well (SIMKOVA, 1966). Chronic infections of Machupo virus in adult Cricetid rodents have been noted, also, with excretion of virus from the urine over long periods (JOHNSON et al., 1965).

There is little evidence to indicate that infectious levels of virus can again circulate in the blood of a bird, rodent or larger mammal once an antibody response has been stimulated. Latent infections of long standing have been detected in birds, however, with virus on occasion isolated from various tissues despite presence of circulating antibody (REEVES et al., 1958).

A different response to infection has been observed experimentally in snakes and bats. In the snakes, viremias of eastern and western encephalitis viruses lasted for long periods of time, frequently in levels known to be infectious for mosquito vector species. If infected in the fall, they carried virus through the winter and continued to circulate it at fluctuating level for a matter of weeks. Antibody responses were sometimes absent, weak, or short lived (KARSTAD, 1961; HAYES et al., 1964; THOMAS et al., 1958; THOMAS and EKLUND, 1960; THOMAS and EKLUND, 1962). A similar pattern of infection has been observed in bats infected with Japanese B encephalitis virus. Very long viremias, weak antibody responses, survival of virus through periods of host hibernation, and periodic recirculation of virus for two to three years, and even transplacental transmission have been reported by Sulkin (Sulkin et al., 1963; Sulkin et al., 1964; Sulkin et al., 1966a; Sulkin et al., 1966b).

Vertebrates as Arbovirus Reservoirs

In temperate regions where the vector population disappears during the winter months, it seems likely that a carrier state in certain vertebrate species serves as a reservoir mechanism to assure continued infection. Birds do not appear to fit well into such a scheme since they have short-term viremias and are usually strong antibody producers. Despite the finding of latent infections in birds, recirculation of virus in infectious levels has not been shown (Reeves et al., 1958). It is more generally believed that they reintroduce virus annually into various temperate areas from tropical zones during their spring migrations. However, the great regularity of recurrence of virus in the same temperate locales, year after year, casts some doubt on the reintroduction theory, which implies an element of chance. Furthermore, regularity of recurrence is not restricted to viruses which infect migrating species. Many viruses which do not infect birds and have no obvious means of reintroduction from afar reappear annually on schedule also. Certainly for them a local reservoir mechanism is indicated. If a local reservoir is possible for them, one for the bird-infecting viruses appears possible as well.

It has been suggested that silent infections in hibernating rodents might serve as the reservoir mechanism. In fact, winter isolations of eastern and western encephalitis viruses from blood or brain tissue of certain rodent species in New Jersey have been reported by Goldfield and Sussman (1964). The viremias detected may have been similar to the extended type reported for hibernating dormice infected with central European tick-borne encephalitis virus (Hoogstraal, 1966), or may have been due to flare-up of old infections. If indeed a recrudescence allowed virus to reappear in the blood in significant level, important physiological changes not yet understood must under some conditions take place in the host to reduce antibody activity or depress antibody already formed. Obviously this area needs more study. For viruses, such as Machupo virus, apparently capable of serial transmission among the vertebrates hosts through the agency of infected urine, neither a viremia nor a vector appears necessary (Johnson et al., 1965).

The findings in snakes suggest that these and other cold-blooded vertebrates might serve as effective reservoirs for some of the arboviruses. Long viremias, ability to hibernate and carry virus through the winter, and recirculation of virus for many days or weeks should adapt them well for this purpose. Bats, also, might serve both as virus reservoirs and to disseminate viruses in view of their long, recurrent viremias, transplacental transmission and migratory habits.

Unfortunately, the importance of these various arbovirus reservoir systems is still questionable. An occassional winter isolation of virus does not necessarily prove that the host species was performing a significant function, and long viremias in snakes and bats could be only interesting biological phenomena without actual importance in virus perpetuation.

Pathogenicity and Symptomatology in Vertebrates

Pathogenicity and symptomatology in vertebrates are not reliable properties of the various arboviruses. In the natural hosts clinical manifestations are usually absent or of an evanescent nature. Birds with St. Louis encephalitis virus infection, for example, rarely show abnormal reaction, and Colorado tick fever virus causes only inapparent infections in its wild rodent hosts. This is apparently true also for the virus of Venezuelan equine encephalitis in its natural rodent hosts, despite a great pathogenicity for a wide variety of the vertebrates.

On the basis of clinical observations alone, it is often impossible to make a definite diagnosis of an arbovirus disease in man. Different arboviruses may produce the same CNS symptoms, as evidenced by the great number with the word "encephalitis" as part of their name, e.g., eastern, western, St. Louis, Japanese B, and California encephalitis viruses. Diagnosis is further complicated by the fact that the encephalitic syndrome represents the extreme disease condition, with subclinical or mild systemic infections usually outnumbering the CNS involved in a ratio of 25—400 to 1. Severe hemorrhagic disease is also produced by at least three arbovirus complexes: Machupo and Junin of the Tacaribe group. Omsk and Kyasanur Forest of the tick-borne complex, and members of the dengue complex (CLARKE and CASALS, 1965; CASALS and CLARKE, 1965b). A relatively mild hemorrhagic disease has also been associated with infections of Chikingunya virus, a member of Group A (CASALS and CLARKE, 1965a).

It is also confusing that different strains of a single virus, or even the same strain, may produce a variety of symptoms in man. Initial infections of arboviruses are systemic, with varying degrees of affinity for different organs or tissues. Most arboviruses are also latently neurotropic. In man, Venezuelan equine encephalitis symptoms range from influenza-like to severe CNS involvement (CASALS and CLARKE, 1965a). The usual symptoms produced by dengue virus strains are severe aches, pains and high fever, yet they may cause the fulminating hemorrhagic disease mentioned above.

Origin of Arboviruses

One might speculate on the origin of arboviruses, whether they arose first in arthropods or vertebrates. Plausible arguments can be presented for both sides, but none have been proven.

A great number of the arboviruses will multiply in many kinds of insects ranging from mosquitoes to moths and mealworms, provided the gut is bypassed and these viruses are inoculated directly into the body cavity (HURLBUT and THOMAS, 1960). This demonstrates a general susceptibility of insect tissues for these viruses and suggests a long-established association between arboviruses and insects. This evidence seems to favor their arising in insects or other arthropods and secondarily becoming adapted to vertebrates, although with a complex of viruses as diversified as the arboviruses, multiple lines of origin are possible.

There are arboviruses which apparently have lost the need for an arthropod vector. An example is U.S. bat salivary gland virus. This virus, on the basis of serological relationship, is a member of Group B which comprises a great list of proven mosquito- and tick-borne viruses, yet arthropod vectors for it are still unknown. Arthropod vectors are obviously unnecessary for its maintenance since bats spread it effectively among themselves by bite. Salivary gland infection in the bat is chronic; active infection in the glands of an individual bat has been proven to last for at least 309 days (BAER and WOODALL, 1966). Possibly in addition to transmission by bats, a basic arthropod-borne cycle has existed or may still exist in association with some other vertebrate group. However, it does not appear to be mosquito borne since it failed to grow in mosquitoes when inoculated intrathoracicly (CLARKE and CASALS, 1965).

Machupo and Junin viruses are examples of viruses which might have arisen in vertebrates rather than arthropods. Unlike the bat salivary gland virus, they show no relationship to a well-documented vector-borne group. Machupo, at least, has been shown to be excreted in urine and is probably thus transmitted from rodent to rodent (JOHNSON et al., 1965). No evidence of arthropod involvement has yet been found in the case of Machupo, but mites have been suspected in the transmission of Junin (CASALS and CLARKE, 1965b). Viremias are present in animals infected with either virus, indicating that vector transmission may be possible. From these clues, it seems likely that Machupo, Junin and the remainder of the Tacaribe group are primarily vertebrate viruses, and if vector borne, only secondarily so.

Encephalomyocarditis virus is an example of a virus which might eventually become vector borne. There seems little need for this, however, since it is transmitted effectively between rodents by contact, apparently through urine contamination as with Machupo. It does not possess some of the characteristics of arboviruses and is not included in arbovirus grouping, yet it has been isolated from mosquitoes on several occasions. Viremias may occur in infected rodents, but attempted experimental transmission by mosquito bite has thus far failed.

Laboratory Virus Detection Systems

Suckling white Swiss mice, one- to three-days-old, inoculated by the intracerebral route, have become the standard test animal for arbovirus isolation since they show such a broad susceptibility to arboviruses in general. They are also commonly used for serum neutralization tests where other systems are not applicable. Presence of virus infection is usually exhibited by paralysis, prostration and death, but with a few viruses, blind passaging may be necessary to build up pathogenicity.

The advantages of using tissue cultures instead of mice for much of the arbovirus work usually center around cost and ease of handling. Unfortunately none of the culture systems currently known have the broad susceptibility and sensitivity of suckling mice, but for a particular virus or combination of viruses, a culture system or group of systems can usually be selected which will serve adequately.

The culture system with the broadest susceptibility spectrum of any thus far tested is a baby hamster kidney cell line, BHK-21. It approaches or equals the susceptibility of suckling mice for a large variety of arboviruses. This cell line, clone 13, has been shown to be susceptible to 129 out of 134 different arboviruses tested, with great enough sensitivity to most of them to appear useful for routine isolation work (KARABATSOS and BUCKLEY, 1967). However, the virus strains tested were laboratory virus stocks with mouse brain passage histories ranging from 2 to >50. Susceptibility to field strains might not be equal in all cases. Other useful cell types are primary hamster kidney, chick embryo, duck embryo, HeLa, L cells, various lines of green monkey kidney including Vero, and porcine kidney, to name several which have been commonly employed. Both fluid cultures and cultures overlaid with agar are useful for arbovirus work, depending upon the types of viruses being studied and the kind of cells used.

Considerable effort in recent years has been made to develop insect tissue cultures, with the thought that they might be highly susceptible to arboviruses. Difficulty has been encountered, however, in establishing and maintaining such cultures since their nutritional requirements are different from those of vertebrate cells and little understood. The most successful operations have required insect plasma in the media, which can be a serious drawback for routine operating procedures.

Theoretically, insect tissue cultures should display ultimate susceptibility to a whole gamut of arboviruses, probably even including groups hitherto unknown. Since arboviruses rarely cause detectable pathology in cells of living arthropod vectors, however, infections in the cell cultures are likely to be inapparent also. If this is the case, indirect methods would be required to determine whether infection actually takes place.

Public Health Importance

About 75 arboviruses are known to cause disease in man (CABV, 1967), and probably many more can produce disease but as yet have been insufficiently studied to establish the fact. It seems appropriate to mention economic importance also since great cost is associated with epidemics, in their prevention, control, investigation and treatment, and through a multitude of various incidental losses incurred.

Any epidemic of arthropod-borne disease causes considerable public concern. Thirty-three cases of eastern encephalitis in New Jersey in 1959 with 21 deaths (MMWR, 1960) were enough to alarm the entire east coast. They inspired much research and stimulated many meetings, ranging from local neighborhood campaigns for mosquito reduction to large conferences involving arbovirus specialists and medical and veterinary authorities.

One of the most serious arbovirus epidemics in the United States in modern times was of St. Louis encephalitis in the Tampa Bay area of Florida in 1962 (BOND et al., 1965). It was centered about St. Petersburg, a popular retirement city. If one imagines a populace overloaded with old people, a disease with a predilection for the older age groups, and 222 cases with 43 deaths, the near panic created can be understood. In addition to the disease problem, real estate values tumbled and the tourist trade slackened. Following this by two years there was another outbreak of St. Louis encephalitis of similar proportions in Houston, Texas (ANON, 1965), and in 1966, others in Dallas and Corpus Christi (MMWR, 1966). The repeated occurrence of major epidemics has made it a predominant public health problem and the most important arbovirus disease in the United States.

Western encephalitis also accounts for many cases annually but these cases occur mainly in rural or semi-rural areas of western United States. California encephalitis, produced by closely related strains in the California arbovirus group, is just beginning to be recognized as an important disease in the United States. The virus is widely distributed, and wherever diligent search has been made for evidence of disease, cases have been found. A considerable number have been reported in Indiana (YOUNG, 1966), Ohio (CRAMBLETT et al., 1966) and Wisconsin (THOMPSON and EVANS, 1965; THOMPSON et al., 1965), and others noted in New Jersey (GOLDFIELD, 1967), North Carolina (MMWR Suppl., 1966) and Florida (QUICK et al., 1965; BOND et al., 1966).

Probably the largest epidemic of an arbovirus disease to occur in the western hemisphere was one of Venezuelan equine encephalitis which swept from Colombia through Venezuelan between 1962 and 1964 (SELLERS et al., 1965). There were 31,966 cases reported with 190 deaths. Reporting was incomplete from many areas, so that the total number of cases must have been much greater. This does not take into account vast losses of horses and burros, upon which the economy of much of the area depends.

Arbovirus epidemics seem to be occurring with greater frequency in many parts of the world. A large outbreak of dengue occurred in the Carribean area in late 1963 into 1964. Many thousands of cases were seen in Jamaica and an estimated 100,000 cases occurred in Puerto Rico (CLARKE and CASALS, 1965). Dengue, endemic in the Philippines and southeast Asia, has assumed new characteristics in those areas in recent years, producing frequently a deadly hemorrhagic disease rather than its usual symptoms. Several large epidemics have occurred, with thousands of cases and a high death rate (HALSTEAD, 1966). Yellow fever flared in Senegal, West Africa, in 1965, demonstrating the danger of relaxing a general vaccination program in a high risk area (BRES, 1966). O'nyong-nyong fever produced a great epidemic in East Africa between 1959—1962 (East African Virus Research Institute, 1962). This disease, recognized for the first time in this outbreak, broke out in Uganda and swept through Kenya, Tanganyika and Nyassaland and affected an estimated 2,000,000 people. Kyasanur Forest Disease was first recognized in Mysore state, India, in 1957 (WORK, 1958). Since that time several hundred cases have been reported with many deaths.

The apparent flare-up of arbovirus disease in various parts of the world may be in part due to better recognition. More investigators than ever before are on the alert for evidence of arbovirus disease. Perhaps also the disease symptoms are becoming more apparant through suppression of malaria, polio and other afflictions which tended to mask them. Part of the increased disease may be related to population growth and extension of human populations into enzootic foci, bringing man into closer association with natural vectors.

Despite the added emphasis on arbovirus disease recognition, a vast number of cases are undoubtedly still going unrecognized. Vague syndromes, fevers of unknown origin and cases of undiagnosed encephalitis occur in almost any area; many of these may be caused by arboviruses. Where effort has been made, it has frequently been possible to associate symptoms such as these with arbovirus infections.

Control of Arbovirus Disease

Vaccination

Ideal control of disease would be through eradication of the causative agents. This is impractical in the case of arboviruses, however, because of their association with wild vertebrates for endemic maintenance. Eradication was once believed possible for yellow fever; now it is realized that the virus is involved with inaccessible jungle reservoirs.

The next best control should be through widespread vaccination with attenuated, live-virus vaccines, one inoculation of which would confer years of immunity. An outstanding example available for human use is the 17 D strain of yellow fever vaccine (THEILER, 1951). It is safe, highly effective and particularly valuable since it protects against an arbovirus which causes a high rate of clinical infection and severe disease. Unfortunately live vaccines for other arboviruses are not available or approved for wide-scale human use. The problems of safety testing and field trials are overwhelming, and considering the large number of people who would need to be vaccinated to protect a relative few from contracting severe cases of such infections as St. Louis encephalitis, it is unlikely that they will be used. Killed vaccines can be prepared, but their efficacy is sometimes doubtful and three inoculations are generally required to provide one or two years of protection. In the event of a disease outbreak in a major city, it may already be too late to vaccinate, and without means of predicting where disease will strike, the problem appears hopeless. Such vaccines are valuable, however, for the protection of laboratory or field personnel who are frequently exposed because of the nature of their work.

Theoretically, the use of vaccines on wild hosts to interfere with build-up of virus infections in nature could be highly effective but has up to now received little thought. A single example which might be applicable is found in current studies at the National Communicable Disease Center on vaccination of wild foxes against rabies by means of special automatic devices (HUDSON, 1967).

Vector Reduction

Vector reduction offers the best means of control currently at hand. It is particularly applicable to mosquito-borne viruses and is done preferably prior to virus build-up to prevent danger altogether. This is best effected by a continued vector control program. In areas where no regular mosquito abatement has been carried out, emergency mosquito adulticide measures are applied as soon as an epidemic is recognized in attempts to eliminate the infected mosquitoes. What appears to have been an abortion of an arbovirus epidemic was demonstrated in Dallas, Texas, in the summer of 1966, when the city was experiencing a large outbreak of St. Louis encephalitis. An aerial mist-spraying of a very low volume of concentrated Malathion insecticide effected an almost complete kill of the adults of the vector species, *Culex pipiens quinquefasciatus*. The vector population remained very low for over a week, presumably time enough for most of the infected host birds in the area to lose their viremias. Prior to the spray treatment, one in every 160 of the mosquitoes tested had been found infected. When a new crop of mosquitoes finally reappeared, there were apparently few viremic birds left to feed upon; the infection rate of the new vector population was only about 1:30,000, far below a dangerous level for transmission to humans (CHAMBERLAIN, 1966; MMWR, 1966).

Vector reduction for control of non-urban disease is more difficult and expensive because of the greater area which must be treated per person protected. Personal measures for prevention of bites by vectors are then heavily stressed, such as the use of mosquito and tick repellents, tick-proof clothing, household insecticide sprays, and protective screening.

Wild Vertebrate Host Reduction

Vertebrate host reduction may in some situations be the most effective means to control an epidemic. This was outstandingly demonstrated in the Bolivian community of San Joaquin, suffering a disastrous epidemic of Machupo virus infection. A classic ecological investigation showed a wild rat species, *Calomys callosus*, to be the vertebrate host. Control of this species promptly ended the epidemic (MACKENZIE, 1965; KUNS, 1965).

Control of urban St. Louis encephalitis through reduction of pest bird species is currently being considered. Normal populations of non-pest wild birds apparently pose no real problem, being neither so concentrated nor so closely associated with man as to be dangerous in most situations. The house sparrow populations have increased so extensively in some cities, however, along with feral pigeons, that they furnish a huge supply of susceptibles in close association with the mosquito vectors and man. Reduction of their numbers only 50 percent would probably in most instances make large-scale human involvement unlikely since epidemics reflect highly abnormal balances which should be relatively easily equalized. Limitation of available food supply can quickly decrease house sparrow populations (LORD, 1967), and an educational program of non-tolerance for pest birds might also inspire application of measures to reduce the numbers of the ubiquitous pigeon.

Control measures thus far considered have been slanted toward interruption of the epidemic cycle. No means of reducing arboviruses appreciably at their wild enzootic source have thus far been demonstrated. If means are found, however, they may well involve manipulation of the vertebrate host population in some manner to make transmission less effective and will probably be restricted to viruses with a limited wild vertebrate host spectrum.

References

ANDREWES, C. H.: Classification of viruses of vertebrates. Advanc. Virus Res. **9**, 271—296 (1962).

—, and D. M. HORSTMANN: The susceptibility of viruses to ethyl ether. J. gen. Microbiol. **3**, 290—297 (1949).

ANONYMOUS: Epidemic St. Louis encephalitis in Houston, 1964, a cooperative study. J. Amer. med. Ass. **193**, 139—146 (1965).

BAER, G. M., and D. F. WOODALL: Bat salivary gland virus carrier state in a naturally infected Mexican freetail bat. Amer. J. trop. Med. Hyg. **15**, 769—771 (1966).

BERGOLD, G. H.: Personal communication 1967.

BOND, J. O., D. T. QUICK, J. J. WITTE, and H. C. OARD: The 1962 epidemic of St. Louis encephalitis in Florida. I. Epidemiologic observations. Amer. J. Epidemiology **81**, 392—404 (1965).

—, W. McD. HAMMON, A. L. LEWIS, G. E. SATHER, and D. J. TAYLOR: California group arboviruses in Florida and report of a new strain, Keystone virus. Public Hth Rep. **81**, 607—613 (1966).

BOORMAN, J.: Observations on the amount of virus present in the hemolymph of *A. aegypti* infected with Uganda S, yellow fever and Semliki Forest Virus. Trans. roy. Soc. trop. Med. Hyg. **54**, 362—365 (1960).

BOWNE, J. G., and R. H. JONES: Observations on blue tongue virus in the salivary glands of an insect vector, *Culicoides variipennis*. Virology **30**, 127—133 (1966).

BRES, P.: L'epidemie de fievre jaune de la region de Diourbel, November-December, 1965 (The epidemic of yellow fever in the region of Diourbel). Sixth Technical Conference of the OCCGE at Babo-Dioulasso, Upper Volta, March 21—26, 1966. (Technical Document No XXIV/8 (1966).

BURTON, A. N., J. MCLINTOCK, and J. G. REMPEL: Western equine encephalitis virus in Saskatchewan garter snakes and leopard frogs. Science **154**, 1029—1031 (1966).

CASALS, J.: The arthropod-borne group of animal viruses. Trans. N.Y. Acad. Sci., Ser. 2, **19**, 219—235 (1957).

— Special points about classification of arboviruses. In: Symp. IX Internat. Congr. for Microbiol. Moscow: Ivanovski Institute of Virology 1966.

—, and L. V. BROWN: Hemagglutination with arthropod-borne viruses. J. exp. Med. **99**, 429—449 (1954).

—, and D. H. CLARKE: Arboviruses; group A. In: Viral and rickettsial diseases of man, fourth ed. F. L. HORSFALL jr., and I. TAMM, ed.). Philadelphia: J. B. Lippincott Co. 1965a.

— — Arboviruses other than groups A and B. In: Viral and rickettsial diseases of man, fourth ed. (F. L. HORSFALL jr. and I. TAMM, ed.). Philadelphia: J. B. Lippincott Co. 1965b.

—, and W. C. REEVES: The arboviruses. In: Viral and rickettsial infections of man, fourth ed. (F. L. HORSFALL jr. and I. TAMM, ed.). Philadelphia: J. B. Lippincott Co. 1965.

CATALOGUE OF ARTHROPOD-BORNE VIRUSES: To be printed, U.S. Govt. Printing Office, Washington, D. C. (1967).

CAUSEY, O. R., R. E. SHOPE, and G. BENSABATH: Marco, timbo and chaco, newly recognized arboviruses from lizards of Brazil. Amer. J. trop. Med. Hyg. **95** (2), 239—243 (1966.)

Chamberlain, R. W.: Unpublished data (1966).
— R. H. Gogel, and W. D. Sudia: Experimental vector studies with strains of St. Louis encephalitis virus isolated from mosquitoes during 1964 epidemics. J. Med. Ent. **3**, 268—270 (1966).
—, and R. K. Sikes: Laboratory investigations on the role of bird mites in the transmission of eastern and western equine encephalitis. Amer. J. trop. Med. Hyg. **4**, 106—118 (1955).
— — D. B. Nelson, and W. D. Sudia: Studies on the North American arthropod-borne encephalitides. VI. Quantitative determinations of virus-vector relationships. Amer. J. Hyg. **60**, 278—285 (1954).
— —, and W. D. Sudia: Attempted laboratory infection of bird mites with the virus of St. Louis encephalitis. Amer. J. trop. Med. Hyg. **6**, 1047—1053 (1957).
—, and W. D. Sudia: The effects of temperature upon the extrinsic incubation of eastern equine encephalitis in mosquitoes. Amer. J. Hyg. **62**, 295—305 (1955).
— — Multiplication of eastern equine encephalitis virus in mosquitoes following intrathoracic inoculation. Proc. Soc. exp. Biol. (N.Y.) **95**, 261—262 (1957).
— — Mechanism of transmission of viruses by mosquitoes. Ann. Rev. Entomol. **6**, 371—390 (1961).
— —, and J. D. Gillett: St. Louis encephalitis virus in mosquitoes. Amer. J. Hyg. **70**, 221—236 (1959).
Clarke, D. H., and J. Casals: Arboviruses; Group B. In: Viral and rickettsial diseases of man, fourth ed. (F. L. Horsfall jr. and I. Tamm, ed.). Philadelphia: J. B. Lippincott Co. 1965.
Cramblett, H. G., H. Stegmiller, and C. Spencer: California encephalitis virus infections in children. Clinical and laboratory studies. J. Amer. med. Ass. **198**, 108—112 (1966).
Davis, N. C.: The effect of various temperatures in modifying the extrinsic incubation period of yellow fever virus in *Aedes aegypti*. Amer. J. Hyg. **16**, 163—176 (1932.
East African Research Institute, Entebbe, Uganda. Report No 12, 1961—62.
Franklin, R. M.: The significance of lipids in animal viruses. An essssay on virus multiplication. Progr. med. Virol. **4**, 1—53 (1962).
Gebhardt, L. P., G. J. Stanton, D. W. Hill, and G. C. Collett: Natural overwintering hosts of the virus of western equine encephalitis. New Engl. J. Med. **271**, 172—177 (1964).
Goldfield, M.: Personal communication 1967.
—, and O. Sussman: Ecologic studies of EE and WE in New Jersey non-avian vertebrate studies. Oral presentation at Communicable Disease Center, U.S. Public Health Service, Atlanta, Ga., August 6, 1964.
Halstead, S. B.: Mosquito-borne haemorrhagic fevers of South and South-east Asia. Bull. Wld. Hlth. Org. **35**, 3—15 (1966).
Hayes, R. O., J. B. Daniels, H. K. Maxfield, and R. E. Wheeler: Field and laboratory studies on eastern encephalitis in warm- and cold-blooded vertebrates. Amer. J. trop. Med. Hyg. **13**, 595—604 (1964).
Hoogstraal, H.: Ticks in relation to human diseases caused by viruses. Ann. Rev. Entomol. **11**, 261—308 (1966).
Howatson, A. F., and G. F. Whitmore: The development and structure of vesicular stomatitis virus. Virology **16**, 466—478 (1962).
Hudson, D. A.: Veterinary Public Healht Section, Epidemiology Program, National Communicable Disease Center. Personal communication 1967.
Hurlbut, H. S., and J. I. Thomas: The experimental host range of the arthropod-borne animal viruses in arthropods. Virology **12**, 391—407 (1960).
Johnson, K. M., R. B. Mackenzie, P. A. Webb, and M. L. Kuns: Chronic infection of rodents by Machupo virus. Science **150**, (3703) 1618—1619 (1965).

JONKERS, A. H., R. E. SHOPE, T. H. G. AITKEN, and L. SPENCE: Cocal virus, a new agent in Trinidad related to vesicular stomatitis virus, type Indiana. Amer. J. vet. Res. **25**, 236—241 (1964).

KARABATSOS, N., and S. M. BUCKLEY: Susceptibility of the baby-hamster kidney-cell line (BHK-21) to infection with arboviruses. Amer. J. trop. Med. Hyg. **16**, 99—105 (1967).

KARSTAD, L.: Reptiles as possible reservoir hosts for eastern encephalitis virus. 26th N. Am. Wildlife Conf., 186—202 (1961).

— O. K. FLETCHER, J. SPALATIN, R. ROBERTS, and R. P. HANSON: Eastern equine encephalomyelitis virus isolated from three species of Diptera from Georgia. Science **125**, 395—396 (1957).

KUNS, M. L.: Epidemiology of Machupo virus infection. II. Ecological and control studies of hemorrhagic fever. Amer. J. trop. Med. Hyg. **14**, 813—816 (1965).

LAMOTTE jr., L. C.: Japanese B encephalitis virus in the organs of infected mosquitoes. Amer. J. Hyg. **72**, 73—87 (1960).

LORD, R. D.: Virology Section, Laboratory Improvement Program, National Communicable Disease Center. Personal communication 1967.

LWOFF, A., R. HORNE, and P. TOURNIER: A system of viruses. Cold Spr. Harb. Symp. quant. Biol. **27**, 51—55 (1962).

MACKENZIE, R. B.: Epidemiology of Machupo virus infection. I. Pattern of human infection, San Joaquin, Bolivia, 1962—1964. Amer. J. trop. Med. Hyg. **14**, 808—813 (1965).

MIMS, C. A., M. F. DAY, and I. D. MARSHALL: Cytopathic effect of Semliki Forest virus in the mosquito, *Aedes aegypti*. Amer. J. trop. Med. Hyg. **15**, 775—784 (1966).

Morbidity and Mortality Weekly Report, Communicable Disease Center, U.S. Department of Health, Education and Welfare, Annual Suppl. **8** (53):2 (September 16, (1960).

Morbidity and Mortality Weekly Report, Communicable Disease Center, U.S. Dept. of Health, Education and Welfare **15**, (39), 334—335 (Oct. 3, 1966).

Morbidity and Mortality Weekly Report, Communicable Disease Center, U.S. Dept. of Health, Education and Welfare, Annual Supplement Summary, 1965, **14** (53): 34 (Oct. 14, 1966).

MURPHY, F. A., P. H. COLEMAN, and S. E. WHITFIELD: Electron microscope observations of Flanders virus. Virology **30**, 315—317 (1966).

MUSSGAY, M., and O. SUAREZ: Multiplication of vesicular stomatitis virus in *Aedes aegypti* L. mosquitoes. Virology **17**, 202—203 (1962).

—, and J. WEIBEL: Electron microscope studies on the development of vesicular stomatitis virus in KB cells. J. Cell. Biol. **16**, 119—129 (1963).

QUICK, D. T., A. G. SMITH, A. L. LEWIS, G. E. SATHER, and W. McD. HAMMON: California encephalitis virus infection: a case report. Amer. J. trop. Med. Hyg. **14**, 456—459 (1965).

REEVES, W. C.: Personal communication. 1967.

— W. M. HAMMON, W. H. DOETCHMAN, H. E. MCCLURE, and G. SATHER: Studies on mites as vectors of Western equine and St. Louis encephalitis virus in California. Amer. J. trop. Med. Hyg. **4**, 90—105 (1955).

— G. A. HUTSON, R. E. BELLAMY, and R. P. SCRIVANI: Chronic latent infections of birds with western encephalomyelitis virus. Proc. Soc. exp. Biol. (N.Y.) **97**, 733—736 (1958).

SELLERS, R. F., G. H. BERGOLD, O. M. SUAREZ, and A. MORALES: Investigations during Venezuelan equine encephalitis outbreaks in Venezuela, 1962—1964. Amer. J. trop. Med. Hyg. **14**, 460—469 (1965).

SHELOKOV, A. I., P. H. PERALTA, and P. GALINDO: Prevalence of human infection with vesicular stomatitis virus. Amer. Soc. clin. Invest. **40**, 1081—1082 (1961).

Simkova, A.: Quantitative study of experimental Tahyna virus infection in hibernating hedgehogs. J. Hyg. Epidem. (Praha) **10**, 499—509 (1966).

Smith, M. G., R. J. Blattner, and F. M. Heys: The isolation of the St. Louis encephalitis virus from chicken mites (*Dermanyssus gallinae*) in nature. Science **100**, 362—363 (1944).

Sudia, W. D.: The multiplication of St. Louis encephalitis virus in two species of mosquitoes: *Culex quinquefasciatus* Say and *Culex pipiens* Linnaeus. Amer. J. Hyg. **70**, 237—245 (1959).

Sulkin, S. E., R. Allen, and R. Sims: Studies of arthropod-borne virus infections in chiroptera. III. Influence of environmental temperature on experimental infection with Japanese B and St. Louis encephalitis viruses. Amer. J. trop. Med. Hyg. **15**, 406—417 (1966a).

— — —, and K. V. Singh: Studies of arthropod-borne virus infections in chiroptera. IV. The immune response of the big brown bat (*Eptesicus f. fuscus*) maintained at various experimental temperatures to experimental Japanese B encephalitis virus infection. Amer. J. trop. Med. Hyg. **15**, 418—427 (1966b).

— R. Sims, and R. Allen: Studies of arthropod-borne virus infections in chiroptera. I. Susceptibility of Insectivorous species to experimental infection with Japanese B and St. Louis encephalitis viruses. Amer. J. trop. Med. Hyg. **12**, 800—814 (1963).

— — — Studies of arthropod-borne virus infections in chiroptera. II. Experiments with Japanese B and St. Louis encephalitis viruses in the gravid bat. Evidence of transplacental transmission. Amer. J. trop. Med. Hyg. **13**, 475—481 (1964).

— C. L. Wisseman jr., E. M. Izumi, and C. Zarafonetis: Mites as possible vectors or reservoirs of equine encephalomyelitis in Texas. Amer. J. trop. Med. Hyg. **4**, 119—135 (1955).

Svehag, S. E., L. Leendertsen, and J. R. Gorham: Sensitivity of bluetongue virus to lipid solvents, trypsin and pH changes and its serological relationship to arboviruses. J. Hyg. (Lond.) **64**, 339—346 (1966).

Theiler, M.: The virus. In: G. K. Strode (ed.), Yellow fever. New York: McGraw-Hill Book Co. 1951.

— Action of sodium desoxycholate on arthropod-borne viruses. Proc. Soc. exp. Biol. (N.Y.) **96**, 380—382 (1957).

Thomas, L. A., and C. M. Eklund: Overwintering of western equine encephalomyelitis in experimentally infected garter snakes and transmissiion to mosquitoes. Proc. Soc. exp. Biol. (N.Y.) **105**, 52—55 (1960).

— — Overwintering of western equine encephalomyelitis virus in garter snakes experimentally infected by *Culex tarsalis*. Proc. Soc. exp. Biol. (N.Y.) **109**, 421—424 (1962).

— —, and W. A. Rush: Susceptibility of garter snakes (*Thamnophis spp.*) to western equine encephalomyelitis virus. Proc. Soc. exp. Biol. (N.Y.) **99**, 698—700 (1958).

Thompson, W. H., and A. S. Evans: California encephalitis virus studies in Wisconsin. Amer. J. Epidem. **81**, 230—244 (1965).

— B. Kalfayan, and R.O. Anslow: Isolation of California encephalitis group virus from a fatal human illness. Amer. J. Epidem. **81**, 245—253 (1965).

Work, T. H.: Russian spring-summer virus in India. Kyasanur Forest disease. Progr. med. Virol. **1**, 248—279 (1958).

Yale Arboviruses Research Unit Annual Report, 1966, permission of the acting director, Dr. Robert Shope.

Young, D. J.: California encephalitis virus. Report of three cases and review of the literature. Ann. intern. Med. **65**, 419—428 (1966).

The Sigma Virus Infection of Drosophila Melanogaster

ROBERT SEECOF*

With 2 Figures

1. History of Research on the Infection

The sigma virus infection of *Drosophila melanogaster* has been the most extensively studied virus infection of multicellular animals from the point of view of genetics. The work has been possible largely because the *Drosophila* host is particularly suitable for laboratory studies, and a great deal of information has been gathered concerning virus transmission patterns and interactions between virus and host genomes. Our present understanding of the infection reveals it to be a hereditary infection, unique in being transmitted by both the egg and sperm of the fly, sometimes in a latent form, sometimes in a virulent form.

The sigma virus infection of Drosophila was discovered 30 years ago by PHILLIPE L'HÉRITIER and GEORGES TEISSIER, who noticed that flies from a certain strain of *D. melanogaster* were permanently paralyzed when exposed to CO_2 in doses that merely anaesthetized other flies of the species. This sensitivity to CO_2 is now known to signify that affected flies carry a sigma-virus infection and CO_2-sensitivity has been shown (KALMUS, 1954) to be ubiquitous in wild populations of *D. melanogaster*.

L'HÉRITIER and TEISSIER reported many of the basic phenomena associated with the infection in an early publication (1945). They found that the CO_2-sensitivity was expressed in infected animals when the treatment atmosphere was kept between —2° C and 23° C and contained 20 to 100 per cent CO_2. Eggs and pupae were not affected but a fraction of the larvae were. The paralysis was apparently specifically induced by CO_2, not merely by anoxia, because virus-free and virus-infected animals responded identically to hydrogen cyanide, carbon monoxide, nitrogen, and other gases. Thoraxes of CO_2-sensitive animals showed paralysis when exposed to CO_2 even when separated from heads or abdomens. The investigators concluded that the infection probably sensitized a carbon dioxide center in the central nervous system, and paralysis then followed exposure to the gas. This is still the best accepted explanation for CO_2-sensitivity.

* Department of Biology, City of Hope Medical Center, Duarte, California, USA.
This work was supported in part by grant AI 05038, U.S. Public Health Service. Contribution 9-67, Department of Biology, City of Hope Medical Center, Duarte, California.

Little additional evidence has been reported concerning the basis for CO_2-sensitivity in the flies. I have found (unpublished results) that virus-infected flies respond identically to injections of sublethal doses of acids, bases, gamma amino butyric acid, adrenaline, D-tubocurarine, physostigmine, and 5-hydroxytryptamine, and flies paralyzed by CO_2 are not relieved by any of these injections. In short, the basis for the CO_2-sensitivity is still uncertain and awaiting the attention of physiologists.

L'HÉRITIER and TEISSIER also reported that CO_2-sensitivity could be inherited, transmitted by eggs, and less efficiently, by sperm. That is, females from the infected strain passed the sensitivity to nearly all of their offspring in every generation regardless of the male parent. When infected males were crossed to noninfected females, however, they passed the sensitivity to only a fraction of their offspring and, of these, only a few of the daughters could transmit it further. The descendents of flies that failed to inherit sensitivity never subsequently manifested sensitivity. Many variations of these basic inheritance patterns have since been reported and will be described later.

The inheritance patterns suggested that a particulate agent (called the sigma particle) was responsible for CO_2-sensitivity. The two investigators, by examining the progeny of infected flies that bore chromosomes marked with visible mutations, showed that the segregation of CO_2-sensitivity was not linked to any chromosome. They concluded the agent was cytoplasmic, but could not decide whether it was a virus or other parasite, or whether it might be some organelle. A virus identity was not attractive because the infection was not transmitted by contact, and infected flies showed no obvious ill effects except for CO_2-sensitivity. Further, the sigma particle was inherited in a regulated fashion, transmitted by sperm, and not readily transmitted by inoculation from infected animals. Therefore, they called it a genoid, a name retained in all of the early reports.

In 1946 and 1947, L'HÉRITIER and HUGON DE SCOEUX reported that CO_2-sensitivity could be transmitted to resistant flies by inoculation with material from affected animals, and inoculated females became transmitters. In addition, an ovary from a resistant fly, when implanted into a sensitive host, produced some sensitivite progeny. These studies established the infectious nature of sigma. It was noteworthy that inoculated male flies behaved like male flies that had received sigma via their father's sperm. They became CO_2-sensitive but not transmitters of the infection. On the other hand, males or females issuing from ovaries that had been infected as implants could become transmitters. Apparently ovaries could be invaded in imagos, but once testes were formed they could not be invaded.

Many studies followed these initial observations on the infection, including descriptions of further responses of infected flies, of the responsible virion, of the effects of temperature and mutation on the life cycle, and hereditary transmission of the virus. In addition, there has been a great deal of speculation concerning facts still undiscovered. The bits and pieces of evidence accumulated have been organized for presentation in Sections 3, 4, and 5. Most of the theoretical treatments that serve to unify these facts have been reserved for Section 6.

2. Responses of infected Flies

A. Preview of the Infection

Details of virus multiplication and hereditary transmission are dealt with in Sections 4 and 5. However, to make this report more understandable, a brief summary of these phenomena will be given before the experimental evidence is presented.

Virus may be introduced to flies by inoculating them with extracts of infected flies, or they may be infected as zygotes by virus carried in sperm. Following introduction by these means, the virus multiplies, producing a yield of infectious material and usually rendering the flies CO_2-sensitive. Infected males will not transmit the infection further, but virus in infected females sometimes invades their developing oocytes and appears in some eggs. Females developing from these virus-infected eggs produce virus which in turn invades their developing oocytes, and in this fashion, a line of flies may be started in which some of the flies are virus-infected and infection stems from oocyte invasion. The infected flies generally produce a yield of infectious material and are CO_2-sensitive. The males do not transmit.

Occasionally, a female arises that apparently has sustained an oogonial infection. She produces progeny that yield infectious material, become CO_2-sensitive, and produce virus-infected eggs and sperm. In this manner, a line of flies can be produced which transmit the virus by maternal inheritance, producing virus-infected eggs which give rise to virus-infected imagos. Flies from these lines originate from infected oogonia in each generation. Any additional virus that invades the oocytes from which they originate, or is carried by sperm into the eggs from which they originate, is largely excluded from multiplication by their prior content of oogonial virus material. When such virus-infected imagos are inoculated with infectious material, the superinfecting virus will not multiply within them, except perhaps very slowly.

A line transmitting the oogonial infection was encountered in which the flies no longer produced infectious material or became CO_2-sensitive. In these flies, the only sign of infection remaining is the immunity to superinfection. This infection, called the ultra-rho condition, may be maintained by the Drosophila as a latent virus virus infection for generations, nondetectable except by the test of superinfection.

A knowledge of Drosophila germ cell formation is necessary in order to interpret the information described in this report. The following account, taken from Demerec (1950), summarizes the events in fly development that are pertinent to germ cell formation.

Fertilization of an egg is followed by a proliferation of nuclei, and their migration throughout the egg before any cell membranes are formed. Cell membranes form after 2—3 hours and then organ formation begins. During organ formation about a dozen cells are maintained apart from the developing embryo in the posterior part of the egg. These cells were formed around nuclei that had migrated there before cell membrane formation and are the primary germ cells. During organ formation these cells undergo a few divisions (to give about 55 cells), and then migrate into the embryo and thence to the gonads.

The adult female ovary contains about 15 ovarioles, each containing an undetermined number of oogonia at its apex. Oogenesis takes place in the ovariole. The details of divisions among the oogonial cells are not well understood. However, cell divisions lead to oocytes that contain nurse cells and are surrounded by follicle cells from the ovariole. Mature eggs have absorbed the nurse cell contents and are fertilized as they are laid by sperm stored in the females. In males, spermatogenesis starts in the larva and the imagos emerge from pupation with a supply of sperm.

B. Mortality and Retardation of Development

Early observations led investigators to conclude that the infection was innocuous to flies, and therefore to suspect that sigma had an identity other than virus. More recent work has revealed that the infection may be damaging to flies throughout their lifespan.

Such damage was first described in my report of 1964 for a strain of flies that had been selected over a number of generations to include a high titer of infectious material in their eggs. About 30 per cent of these eggs failed to yield imagos, and as deaths in sibships were proportional to embryo titers, about 20 per cent of the deaths could be attributed to virus multiplication. Counts of eggs that failed to hatch indicated that nearly all of the deaths occured before egg hatch. The larval or pupal stages were affected too, in that the developmental time required between egg and adult was noticeably increased.

In a later study (1966), I measured the survival of progeny from flies that contained less infectious material than did those of the first strain, but produced a higher percentage of infected eggs. No deaths attributable to virus were detected in this second strain. I concluded that measurements indicating that the second strain contained relatively little infectious material were true. This confirmed the efficiency of the standard assay technique. The efficient infection of eggs by this strain was probably not attributable to a large virus population overlooked by the assay but to some other mechanism of regulation.

C. Lowering of Imaginal Fecundity.

Jupin (personal communication) found that females lay fewer eggs after inoculation with certain virus strains. Inoculated females laid normally fertile eggs until the time virus invaded their oocytes. Thereafter, nearly mature eggs (stage 14, KING et al., 1956) accumulated in their ovaries but few were laid. Jupin suggested that sigma infection of an oocyte slows its development. While this is probably true, the possibility remains that oocyte development fails because of virus activity elsewhere in the fly about the time of egg invasion. No detailed study concerning the effect of inherited virus on fecundity has been reported.

D. CO_2-Sensitivity of the Imago

CO_2-sensitivity is, of course, another response by the flies to the infection. An incubation period following the initiation of infection is always required before the sensitivity appears. Usually this period is correlated with the incubation time during which the virus is increasing to its maximum titer in the fly, and both the maximum titer and the CO_2-sensitivity appear at about the same time (PLUS, 1954; SEECOF, 1962; BRUN, 1963). Arbitrarily, a fly is considered CO_2-sensitive when it fails to recover in 15 minutes at 25° C from a 15-minute exposure to 100 percent CO_2 at 15° C (L'HERITIER and SIGOT, 1946). There is a short period of time just before a fly is fully CO_2-sensitive in which it may recover in very nearly 15 minutes, making classification difficult. However, because such flies are seldom encountered, the test is essentially unequivocal.

We know that CO_2-sensitivity is not related simply to the titer of infectious material in the animal. Some fly strains manifest CO_2-sensitivity when the titer reaches about 10^4 infectious units (IU) per fly (PLUS, 1954). Other combinations of fly and virus genotypes are known in which few IU are formed in the flies and they may not become CO_2-sensitive at all (OHANESSIAN, 1963; BRUN, 1963; VIGIER, 1966). These facts suggest that the infection must somehow progress to a certain point in virion production before CO_2-sensitivity is reached. On the other hand, many virus strains induce CO_2-sensitivity although they fail to reach high titers in the flies (SEECOF, 1962; OHANESSIAN, 1963). Moreover, flies which acquire the infection late in imaginal life may maintain high titers for a period of time before becoming CO_2-sensitive (SEECOF, 1965). A high titer of infectious material is therefore not a sufficient condition for CO_2-sensitivity. Perhaps a certain organ must be penetrated or a certain humoral factor elaborated before CO_2-sensitivity is manifested.

E. Assay for Sigma

Although the appearance of CO_2-sensitivity is affected by the abovementioned factors, the appearance of sensitivity in inoculated flies is a positive identification of sigma in the inoculum. This fact serves as the basis for an assay for sigma based on end-point dilution. PLUS (1954) inoculated serial dilutions of sigma into groups of virus-free flies, then tested them for CO_2-sensitivity at intervals. She found, as expected, that only a fraction of the flies in groups that received dilute inocula became CO_2-sensitive. These sensitives were distributed among the groups of flies in Poisson fashion, with frequencies indicating that a single IU was sufficient to render a fly CO_2-sensitive. A single IU was found sufficient for infection for other sigma strains too, when these were later tested (SEECOF, 1962; BRUN, 1963; OHANESSIAN-GUILLEMAIN, 1963).

Moreover, PLUS found that the incubation time preceding the appearance of sensitivity in inoculated flies was inversely proportional to the logarithm of the number of IU injected. Thus, flies receiving many IU became sensitive in a matter of days; those receiving few IU, in weeks. In this fashion, a standard curve was drawn which related the incubation time for sensitivity to the inoculum size, and the curve was used for a simplified assay.

Rather than make end-point dilutions investigators now usually titer an extract by inoculating a standard volume into a group of flies of standard genotype. Then they subject the flies to CO_2 daily, determine the average time required for sensitivity to appear, and convert this time to a titer by means of a standard curve made for that virus strain under identical conditions. Of course, the reference strain flies used to test the inocula need to be chosen carefully. They should not contain genes such as those described below, which lead to the inhibition of virus multiplication. They must also be uniformly young, as imaginal resistance to virus multiplication increases with age (BRUN, 1963; SEECOF, 1965).

The sensitivity of this assay is limited usually by the volume of the inoculum delivered during injection. A recipient fly cannot tolerate more than about 0.5 μl. Even if this much is delivered, the recipient flies will not receive more than 1 IU on the average unless the initial titer of the inoculum is more than 2×10^3 IU per ml. However, if the sample to be assayed is homogenized in a very small volume, all of the extract can be injected into test flies and titers close to zero can be measured. Throughout the present report, the figures given for titers refer to extractable infectious material titered by this assay.

3. Description of the Viron

A. Physical Characteristics

Recently, electron micrographs were taken of a virus in flies which is almost certainly the causative agent of CO_2-sensitivity. BERKALOFF et al. (1965), studying a sibship of 18 females, detected virions in all of the 10 females known to be virus-infected; and none were found in the eight virus-free females.

Ovarian tissue was examined in section and homogenates of whole flies were examined on grids. Virions were found in both types of preparations but only in low number. Those in the ovary were located in the spaces between follicle cells. The virion was a rod 70×140 mμ in dimension, hemispherical at one end and blunt at the other.

The virion morphology agrees very well with the physical characteristics previously attributed to it by less direct measurements. L'HÉRITIER and PLUS (1950) had calculated a diameter of 42 mμ for the volume sensitive to X-ray inactivation and a diameter of about 180 mμ from filtration data. The morphology and dimensions reported for sigma virus also agree very closely with those for vesicular stomatitis virus. HAWATSON and WHITMORE (1962) reported that virus to have a similar shape and dimensions of 68×175 mμ.

B. Methods of Purification

Several systematic attempts to purify and concentrate the sigma virus have been reported. HUGON DE SCOEUX (1952) and PLUS (1960) explored several buffers, pH ranges, dialysis procedures, and differential centrifugation routines. The most successful procedure, developed by PLUS, calls for homogenization of the flies in 0.25 M sucrose containing 0.00075 M "Tris" buffer at pH 7.45, followed by sedimentation between 6,000 and 20,000×g. This resulted in a purification of about 20 on a protein basis and a yield of close to 100 per cent.

Later, PLUS (1962) centrifuged extracts containing virus in a sucrose gradient of 0—60 per cent. This procedure was not very useful because sucrose concentrations above 40 per cent were found to inactivate the virus. However, PLUS found that the infectivity tended to disperse with the particulates in suspension rather than remain as a discrete band. Virus from a single level in the gradient was used as an inoculum in flies and the progeny virus recentrifuged. This virus displayed the same apparent heterogeneity. PLUS concluded that the heterogeneity was not inherent in the virions but rather followed from their association with the suspension particulates.

An extract was purified by PLUS's differential centrifugation procedure, then centrifuged to equilibrium in CeCl (SEECOF, 1962). A peak of infectivity containing about 10^7 infectious units per ml was obtained at density 1.24. This fraction was accompanied by a visible band of contaminating particulates which later also proved to be present in extracts of virus-free flies.

The association of infectivity and cell particulates, demonstrated in both the sucrose and CeCl gradients, is probably responsible for the failure to further purify the virus. Suggestively, a similar phenomenon has been reported for extracts containing vesicular stomatitis virus. HACKETT (1964) found that this virus could not be purified readily by differential centrifugation because it adhered strongly to membranous structures.

C. Nucleic Acid Content

The identity of sigma's nucleic acid is still conjectural. Morphologically the virion resembles that of vesicular stomatitis or rabies virus, both RNA viruses. As no viruses with this morphology have been found to contain DNA, one might speculate that sigma is an RNA virus.

PLUS (1963) reported that 5-fluorodeoxyuridine, an inhibitor of DNA synthesis, delays the multiplication of sigma virus in flies. The effect of the analog, however, was not reversed by thymidine as would have been expected if the analog were blocking DNA synthesis by its inhibitory effect on thymidine synthetase. Thymidine, in fact, added to the inhibition of sigma multiplication. The failure of the reversal could be interpreted to mean that the analog did not interfere with sigma multiplication by halting DNA synthesis. PLUS, however, showed that deoxycytidine reversed the analog and the thymidine blocks. She suggested that both the analog and the thymidine acted to block deoxycytidine formation and, as deoxycitidine was a DNA precursor, that sigma virus probably contained DNA.

So far it has not been possible to obtain a preparation of sigma virus purified and concentrated sufficiently to allow direct chemical analysis. Neither has there been any report of the successful labeling of the virus with any radioactive precursor molecule.

4. Evidence Concerning Virus Multiplication

A. Growth Curves

Inoculation of sigma into virus-free larvae, pupae, or imagos (PLUS, 1954) is usually followed by multiplication of the virus. Typical curves for multiplication in inoculated imagos were presented by PLUS (1951, 1954, 1955), who inoculated a large number of flies and sacrificed some at intervals to determine their content of infectious material. An initial decrease of IU, which probably represented the eclipse phase of the virus, was followed by an exponential increase in titer which probably represented virion formation.

The titer was found to reach a final maximum value and plateau there for the remainder of the flies' lifetime. BRUN (1963) showed later that the final plateau level of titer depended on the genotype of the inoculated virus, not on the size of the inoculum. LESTRANGE (1963) concluded from her study of temperature effects, reviewed below, that the plateau represented a dynamic equilibrium of particle formation and destruction rather than a cessation of virus production.

For flies that inherit the infection because they develop from infected oogonia, virus multiplication starts in the oocytes from which they are derived, continues through fertilization and development, and finally attains the plateau of maximum titer about the time the imago ecloses. BREGLIANO (1965) found, for a certain strain of such flies, that titers increased from less than 10 IU in germaria (which contain an uncertain number of oogonia) to a maximum of 1,000—10,000 IU in stage 10B cysts and then dropped to 50—1,000 IU in cysts containing nearly mature eggs. (see KING et al., 1956, for a description of the stages of oogenesis.) About 20 per cent of the infectious material in stage 10 B cysts was attributed to the oocyte proper. The fraction of IU within the oocytes at other stages was not reported.

In an earlier study, BREGLIANO (1963) measured virus multiplication for the embryonic and later stages of development. This group displayed about 5—50 IU per early embryo. This increased exponentially to about 10^4 in third instar larvae, decreased to about 10^2 in early pupae, and increased again in late pupae and very young imagos to reach a final plateau of about 10^4 per imago. BREGLIANO attributed the pupal losses to the autolytic processes at that time.

Fig. 1 illustrates the course of virus multiplication in flies issuing from the cross of females maintaining an oogonial infection and virus-free males. The curve, drawn from a composite of data from PLUS (1954) and BREGLIANO, shows the embryos beginning with an appreciable quantity of infectious material inherited, of course, from the oogonial stage.

PLUS also examined virus multiplication in progeny from the reciprocal cross, that is, in offspring of virus-free females that had been crossed to (transmitting) males issuing from an oogonial infection. This curve, as can be seen in Fig. 1, is not similar to that from the first cross. Very few IU could be detected in the embryo, as if only or very few IU had been transmitted by the sperm. Furthermore, the increase in titer was faster in these larvae than in those inheriting virus from the female parent, and progressed to an imaginal plateau about five times as high. No pupal drop in titer was reported, but this may have been overlooked.

The difference in the growth curves of these reciprocal crosses has figured prominently in analyses of the sigma virus infection. Virus multiplies faster and

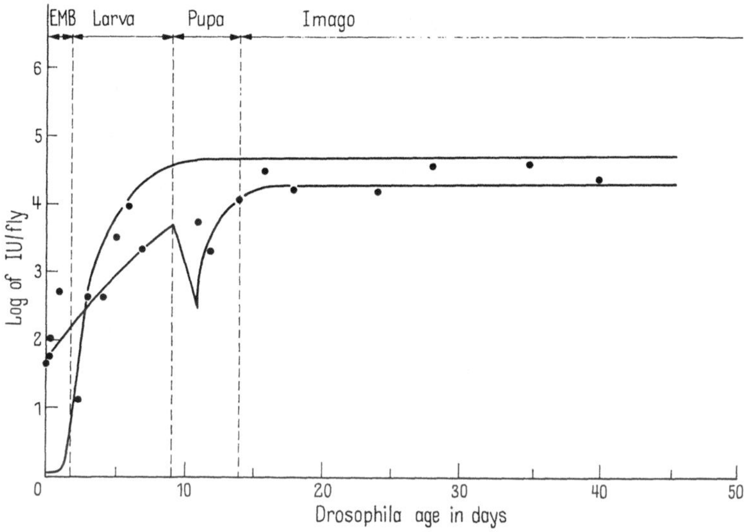

Fig. 1. Formation of infectious material during fly development. Curve 1 is for the progeny of females transmitting an oogonial infection when crossed to virus-free males. Curve 2 is for the progeny of virus-free females crossed to transmitting males. Based on data by PLUS (1954) and BREGLIANO (1963)

to higher titers when inherited via sperm than via eggs infected as oogonia. Oogonial infection seems to confer some kind of resistance to the animal so that virus multiplication is slowed. This difference is probably not due to variability in the fly genotype because all of them were from one inbred strain in the reciprocal crosses made by PLUS. Nor, apparently, is the difference due to virus of a faster growing genotype segregating to sperm. Virus extracted from the progeny of PLUS's reciprocal crosses grew equally well when inoculated into virus-free flies. Whenever this type of reciprocal cross has been re-examined by this technique, a similar result has been obtained (LESTRANGE, 1963; VIGIER, 1966).

In 1962, I showed that imagos maintaining an oogonial infection were also resistant to the multiplication of any additional virus introduced by inoculation. Multiplication of the superinfecting virus was very low or absent. Imagos that were infected by sperm may be resistent too, but this has not been tested.

In the same report, I showed that the low titer exhibited by imagos maintaining an oogonial infection was probably real. That is, no large component of virus was overlooked in the assay because it was slow growing or did not induce CO_2-sensitivity. These conclusions followed from testing extracts of the flies that were diluted until the infectious content was no longer present. Successive blind passages of the diluted material failed to bring any slow-growing virus to a detectable level, and the diluted material failed to induce immunity to superinfection.

BREGLIANO (1965) examined virus multiplication in oocytes from inoculated females. These oocytes originated as virus-free oogonia which were invaded during maturation. The virus was found to multiply in oogenesis in a fashion similar to that reviewed earlier for females with an oogonial infection. Some cysts were found that yielded as many IU as for the latter females but many were found at each stage examined that yielded fewer.

The virus multiplication that followed in the flies developing from these oocytes was not reported by BREGLIANO. Consequently, no information is available as to whether virus multiplication was subsequently rapid, like sperm-initiated infection, or slow, like oogonial-initiated infection. This information could be critical for an understanding of the sigma infection. If resistance to superinfection depends on infection of the somatic cells during embryogenesis, as will be argued later, the average rate of virus multiplication should be slower than in the sperm-initiated infection. If resistance requires prior oogonial infection, it should be faster because of its head-start.

B. Influence of Fly and Virus Mutations

OHANESSIAN-GUILLEMAIN (1959, 1963) and GUILLEMAIN and PLUS (1953) described several fly mutations which lend the animal resistance to sigma infection in that inoculated flies of these genotypes took longer to manifest CO_2-sensitivity than the standard wild type. These genes were located on the first, second, and third chromosomes. The best studied of these genes has been the one called "refractory" (*re*), a semidominant located close to the gene for "black" on the second chromosome (OHANESSIAN-GUILLEMAIN, 1963; OZOLINS, personal communication).

When sigma is inoculated into imagos heterozygous for *re*, the curve for virus multiplication is similar to that in wild-type flies but the titer rise is less than exponential and the final plateau of maximum titer is reduced by about 90 per cent. GUILLEMAIN could show that one IU was still sufficient for infection but the probability of any particle initiating infection was reduced. If the inoculated flies are homozygous for *re*, virus multiplication is prevented altogether.

If the *re* gene is introduced by sperm into re^+ eggs that already contain oogonially-derived virus, the resultant fly is heterozygous for *re*. These animals accumulate infectious material at the same rate as do re^+/re^+ flies. If OHANESSIAN-GUILLEMAIN's interpretation that the *re* gene limits virus nucleic acid synthesis is correct, this means the animals' virus nucleic acid was assured before fertilization. The *re/re*+ females, however, then produce an increased number of virus-free eggs. This would be expected if virus nucleic acid synthesis were decreased in their dividing oogonia.

When *re/re*+ females are mated to *re/re* males, their *re/re* progeny contain little or no infectious material. This is expected if the progeny arose from oogonia that had become reduced in virus content, and virus nucleic acid synthesis were further inhibited by the *re/re* genotype. These *re/re* progeny may contain as few as 50 IU per fly and still be CO_2-sensitive, but some of them are CO_2-resistant. Despite their low titers, all of them produce virus-infected eggs with no less frequency than their heterozygous *re/re*+ mothers, according to Ozolins. This is entirely unexpected. Their eggs presumably were derived from cells containing less virus than in the previous generation and more virus-free eggs should have appeared. In fact, OZOLINS found that their oocytes indeed contained less infectious material and suggested that noninfectious virus nucleic acid accounted for the oocyte infection. This explanation is insufficient because one needs to postulate that the *re* gene inhibits virus nucleic acid synthesis in order to account for the virus-free eggs from *re/re*+ females. The facts seem to argue that some special mechanisms exist to insure that the germ line infection is maintained. Virus nucleic acid or some other form of noninfectious virus seems to go preferentially to oogonia even though the accompanying somatic tissue is reduced in virus content.

If the *re/re* females which produce low-titer oocytes are crossed to *re*+/*re*+ males, the virus grows rapidly to high titers in the progeny. The progenies' germ cells develop free of virus as if the infection had been initiated by sperm or oocyte invasion, even though they stemmed from infected oogonia. These *re/re*+ progeny behave, in other words, as if the oogonia in the *re/re* females from which they originated were so depleted of virus that little virus was present in the oocytes derived from them.

Several phenotypic differences among virus strains have been described (DUHAMEL, 1954; OHANESSIAN-GUILLEMAIN, 1959, 1963. Among these are differential responses to temperature or host genotype, differences in rates of virus multiplication, and differences in the manner by which the virus affects the germ line.

A virus strain may remain stable for a particular character. For example, the ultra-rho behavior described below has never been known to revert to rho behavior. Alternately, virus strains apparently pure for a virus character may accumulate particles changed for that character at a relatively great rate. VIGIER (1966) summarized data to show that most of the virus phenotypes can change at a rate which, if expressed as mutations per particle duplication, would be one in every 100—10,000 particle duplications. This is very high compared to mutation rates calculated for bacterial-virus genes, which generally do not mutate more often than once every 10^8 particle duplications. Curiously, for most of the virus characters, the character may be labile in some virus strains, but may be so stable in others that no changed particles are generally observed.

No basis has been discovered for the rapid changes in virus type observed when characters are labile. However, the virus characters are transmitted as if they reflect differences in the virus genome, albeit polygenic ones, and have been considered to reflect virus mutations in the broadest sense. A great deal of information has been accumulated concerning each virus character, generally in strains where the character is relatively stable. This section is devoted to describing those virus phenotypes characterized by differential rates of virus multiplication. Virus phenotypes concerned with virus transmission will be discussed in a separate section.

Sigma strains which multiply at normal rates despite high temperature are called temperature resistant (Tr). Some virus strains will not render inoculated imagos CO_2-sensitive if the flies are kept at 30° C for the 10 days following inoculation. These strains are called temperature sensitive (Ts) (OHANESSIAN-GUILLEMAIN, 1959).

Certain virus strains can overcome the inhibitory effects of the *re* fly genotype. Viruses which can multiply in the Paris strain of flies, which is homozygous for an *re* gene, are called P^+ to distinguish them from P^- viruses which cannot do so (GUILLEMAIN, 1953).

Virus strains differ too in the time required to render flies CO_2-sensitive following inoculation (BRUN, 1963). For virus strains called "strong", 1 IU may do so in about 15 days, and for strains called "weak", 1 IU may require more than 25 days. The incubation times reflect the rate that infectious material is produced. Strong strains produce high titers relatively quickly and weak strains produce lower final titers during long incubation times.

BRUN showed that the incubation time was a reliable characteristic for defining virus strains and proposed a standard procedure for measuring it. This was to inoculate a female with 1 IU, allow time for the viruses' characteristic final plateau value to be reached, homogenize the female in 0.1 ml and make inoculations with the extract. The average incubation time associated with this extract characterized the virus strain. BRUN called this time (in days) the strain's "dimensional type" or "plaque size", an allusion to the measure of phage yield when a single bacteriophage generates a plaque. He noted that the plaque size was a reflection of the overall growth rate of the virus and probably reflected many virus genes.

Most of the virus strains BRUN examined showed plaque sizes between 9 and 24 days and a continuous spectrum of sizes was detected among virus strains. Examinations of the responses of incubation times to inoculum dose showed that the plaque size character followed from the virus yield per cell rather than the efficacy with which virus strains induced CO_2-sensitivity. As might have been expected, when mixtures of plaque sizes were inoculated into flies, the small plaque sizes (fast growers) outgrew the large plaque sizes.

VIGIER (1966) used a modification of the procedure to measure the virus plaque type. BRUN had shown that each plaque type led to a characteristic plateau titer regardless of the inoculum size. VIGIER, therefore, inoculated one or more flies with inocula of unknown titer, as convenience dictated, instead of

1 female with 1 IU. The two procedures gave equivalent results when clones of known plaque types were used.

VIGIER (1961, 1966) followed the multiplication of plaque size types of virus in flies whose infection was derived from infected oogonia. He assayed the embryos and the imagos within sublines of flies for the proportions of plaque types to be found in each stage. Fast-growing types were found in higher proportion in the imagos than in the embryos, suggesting to him that during fly development some virus multiplication had taken place in which strong virus type had a selective advantage. He argued that some intercellular spread of the infection was necessary to allow this selection. This point is important to considerations of the infection in flies that inherit virus. It is pertinent here to state that his argument may be wrong. It seems possible that strong virus could be selected for during intracellular multiplication.

VIGIER showed further that given females transmitted different proportions of virus types to their offspring than they displayed in their own tissues, and concluded that a flies' somatic content played no role in the germ line transmission of the infection. This conclusion does not follow necessarily from the data either. As BRUN (1963) pointed out, the germ line content may represent only a small sample from the somatic content and therefore need not reflect it accurately.

The virus variant called rho has been described by BRUN (1963). Rho produces infectious material very slowly in flies which inherit it. The infectious material is typical sigma virus in that, once formed, it multiplies according to its plaque-size character in ordinary fashion, and as an inoculum it initiates only sigma and not rho infections. Rho-carrying flies sometimes become CO_2-sensitive very late in life, in association with the formation of infectious material.

BRUN presented a great deal of evidence to show that rho virus is a variant of sigma, not a second factor. He noted that rho-carrying flies arose from sigma-infected ones, and outcrosses showed the noninfectious character of rho to follow from the genotype of the virus, not the fly. The fly gene *re* limited the multiplication of rho as it did sigma virus. That is, the offspring of rho-containing flies of genotype re/re^+ became CO_2-sensitive at an increased rate. The rho infection also responded to heat treatments as if it were a sigma infection. (The heat treatments will be detailed in the following section.) BRUN concluded that the rho variant is a defective virus, incapable of completing virion synthesis, and that infectious material was formed only after virus mutation during the course of fly development.

Flies transmitting a rho infection gave rise to strains called ultra-rho. In ultra-rho strains BRUN (1963) found that infectious material or CO_2-sensitivity never appear during the flies' lifetimes. The only sign of infection remaining in these flies is their resistance to the multiplication of superinfecting virus, should any be inoculated.

Several tests showed that the ultra-rho infection is caused by sigma virus. Ultra-rho-containing flies, like those containing sigma, showed a reduced immunity to superinfection when made heterozygous for the *re* gene and produced some virus-free progeny. Heat treatments affected ultra-rho-bearing flies as it did sigma-infected flies by freeing some of their germ cells of the infection (details below.)

BRUN also argued that ultra-rho-bearing flies were not merely flies bearing a rho variant characterized by a large plaque-size (slow-growing). He noted that neither infectious material nor CO_2-sensitivity ever appeared in flies carrying ultra-rho, spontaneously or in response to heat treatments. A "weak" rho strain would probably have been detected by such signs during BRUN's study, as he tested about 40,000 flies for CO_2-sensitivity. He concluded that the ultra-rho factor was a variant of sigma and probably homozygous for the mutation to rho, that it was unable to initiate virus multiplication and, in addition, it carried the rho trait which prevented the completion of virion synthesis. This hypothesis accounted for all of the data and also for the fact that rho or sigma revertants have never been recovered. Both reverse mutations would have to occur in the same virus genome in a fly in order to yield detectable virions, and this would be a very rare event. It should be pointed out here that one could account for the data by postulating only that the mutation to ultra-rho blocks virion synthesis at an earlier step than does the mutation to rho virus.

C. Influence of Temperature

Some of the earliest observations on the sigma infection were concerned with the effects of heat treatment on the subsequent appearance of CO_2-sensitivity. L'HÉRITIER and SIGOT showed a variety of effects, summarized in a report in 1946. They found that heat treatments of varying duration at about 30° C sometimes "cured" the flies so that the imagos were no longer CO_2-sensitive but that under some conditions the imagos or their offspring could return to sensitivity at lower temperatures. Surprisingly, they found that heat treatment of embryos seemed to prevent heat treatments at a later stage from "curing" the animals. They showed too that heat treatments to maturing gametes permanently cured some of them because these gametes led to progeny that were free of infection.

The significance of these observations will become clear when the later work of LESTRANGE and BRUN is reviewed. These workers found similar effects but they measured titers instead of relying on the appearance of CO_2-sensitivity, so their interpretations are more detailed.

As noted, some sigma variants (Ts) are prevented from rendering their hosts CO_2-sensitive at elevated temperatures. LESTRANGE (1954, 1963) subjected flies carrying Ts virus to temperature changes and measured the changes in titer.

In an initial series of experiments, she heated infected imagos at 30.5° C for several weeks, sampling their titers at intervals. These measurements were made separately for imagos whose infection originated in their mother's oogonia, for imagos infected by their father's sperm, and for imagos that had been infected by inoculation and had already reached their plateau of maximum titer.

All three kinds of infection yielded roughly similar results. Titer dropped exponentially during the first one to two weeks, from about 10^4 IU per fly to 10^2—10^3 IU per fly, then decreased at a slower exponential rate for the remaining weeks of the experiment. The two slopes probably did not represent a rapid decrease in a majority content of Ts virus at first, then a slower decrease in remaining Tr virus, because virus remaining after the presumed Ts virus was eliminated by heating, giving a typical curve with two slopes when retested. The initial drop in flies infected as oogonia or by inoculation (about 0.5 log units per day) was somewhat faster than in flies infected from sperm (about 0.26 log units per day).

In other experiments, LESTRANGE found that heating pupae that had received virus from either parent caused a very rapid drop in titer. As BREGLIANO (1965) later noted, however, sigma titer usually drops during pupation and so the rapidity of the drop was not due to heating alone but possibly to enzymatic degradation as well. Heating inoculated imagos before they had reached their plateau of maximum titer caused their titers to drop during the 48-hour treatment at rates similar to those observed when heating "plateaued" imagos.

In a third series of experiments, LESTRANGE repeated the format of the initial series but this time returned samples of flies to 20° C at intervals during the heat treatment. At 20° C, titer levels rose very rapidly to the level of controls that were never heated. The imaginal titers gave the same response for each of the three different origins of infection. The only differences detected were that the titers in flies that had received virus from their female parents rose slightly higher than control levels, and those from their male parents slightly lower.

Heat treatment to preimaginal stages, oogonium, embryo or larva, was followed by a rapid increase in titer upon a return to 20° C. In embryos, the final titer was higher than controls. The only instance which gave other than a rapid titer return at 20° C was that found in flies inheriting virus and subjected to a prolonged treatment of 30.5° C that lasted from the prepupal stage to the fourteenth day of imaginal life. These required about two weeks to attain high titers. No treatment in any experiment completely freed the animals of virus.

In order to determine whether the increase in virus titer following a return to 20° C was due to virus multiplication or to reactivation, LESTRANGE heated infected flies until their titer was reduced and then inoculated them with heat-resistant virus. These flies were kept at 30.5° C for a period of time and then assayed to determine if the heat treatment had freed the flies of original virus and allowed the superinfecting virus to multiply, or if a noninfectious virus form remained and prevented that multiplication. In these trials, the Tr superinfecting virus also bore the virus character g^- to distinguish it from the Ts g^+ original virus. Any virus finally recovered after the heat treatments was thereby distinguishable. (The virus marker g will be discussed later.)

The superinfection test was made first on imagos. Imagos that had received the infection from their male or female parents or via inoculation all responded the same way; the superinfecting virus failed to multiply. The few superinfecting particles that were recovered were considered to be original inoculum particles.

Secondly, imagos that had received the infection from their male or female parent, but had been heat-treated as pupae, were challenged by superinfection. The superinfecting virus apparently could multiply in these imagos because some of them were positive for virus at a time when the original inoculum was considered to be not detectable. The final titers in the positive superinfected flies (less than 800 per fly) were considered low, but the control titers were not given so the progress of the superinfection is hard to judge.

Surveying these data, LESTRANGE concluded that the drop in titer induced by heating did not represent a permanent inactivation of virus. She considered that the titer increase following a return to 20° C was too rapid to be accounted for by multiplication and must represent reactivation. The one instance of slow titer increase, that following prolonged pupal and imaginal heating, she thought probably followed some permanent virus inactivation. In support of reactivation, she noted that the resistance of "heat cured" flies to superinfection probably meant that a noninfectious form of virus still remained in these flies, ready for reactivation. The limited yield of superinfecting virus following pupal heating she attributed to a few cells that had been entirely freed of the initially infecting virus. Furthermore, LESTRANGE cited personal communications to show that FUDR did not inhibit the titer increases following return to 20° C. If PLUS's results mean that sigma is a DNA virus, this fact suggests that viable sigma nucleic acid remained after heat treatment.

LESTRANGE rejected the possibility that this reactivation followed from a dynamic equilibrium of synthesis and destruction of sigma which ordinarily held within flies. She noted that inoculated sigma invaded its host's eggs only at the onset of the plateau of maximum virus titer, suggesting that only one "crop" of virus was usually made. Furthermore, if the plateau were usually such an equilibrium, FUDR should cause it to decline, and it did not. Another result mitigating against a fixed equilibrium was that virus titers returned to levels somewhat different than the control levels following some heat treatments.

The explanation preferred by LESTRANGE was that infectious virus was usually in dynamic equilibrium with noninfectious virus that had suffered a changed (or lost) protein coat; thus, heat treatment favored noninfectious coats. The coats were restored to their control levels at 20° C, and virus nucleic acid was not damaged except by prolonged heating. With this model LESTRANGE could explain most of the results. The model avoids the objections listed for the first two because only one crop of virus nucleic acid need be postulated, and the virus titer could rise rapidly following return to 20° C because virus nucleic acid did not need to be resynthesized completely.

In summary, it should be noted that none of the models are completely eliminated by the data. The rapid titer rises could be due largely to new virus synthesis, the FUDR data would be irrelevant if sigma finally proves to be an RNA virus, and the arguments concerning one crop of virus before heat treatment are not compelling. Even the resistance to superinfection, which strongly suggests that virus information persists in the cells despite heating, is difficult to use for model-building because the origin of the resistance is unknown.

Nevertheless, though the data are not completely explained, three of LESTRANGE'S considerations are important to the discussion. First, all of the data suggest that some virus can persist during heat treatment in a form which is noninfectious. This form is likely to be virus nucleic acid. Secondly, the data suggest that a content of this noninfectious material lends cells some resistance to superinfection. Thirdly, the data indicated that flies may become mosaics on the cellular level, having some cells free of virus but capable of yielding infectious material (virions) when later infected.

This last conclusion followed from the fact that heat treatments during stages when fly cell division was prominent (heating embryos or heating pupae, then superinfecting the imagos) produced higher titers than the nonheated controls. Presumably heating during division led to some virus-free daughter cells which later supported virion production better than did the cells which retained noninfectious material.

A few results from the heat treatment experiments seem to defy explanation. For example, the titer of flies infected as oogonia dropped faster during heat treatment than the titer of flies infected via sperm. LESTRANGE could only propose that the difference followed from a difference in virus coat stability between the kinds of infections. However, the titer of flies that were infected by inoculation fell at the relatively rapid rate of the oogonially-derived infection instead of falling at the rate of the sperm-derived infection which it usually resembles.

BRUN (1963) applied temperatures of 30° C to developing flies which had inherited a rho or ultra-rho infection from their female parents. Heating as larvae caused the imagos to develop sensitivity at an increased rate both spontaneously or in response to a superinfection by sigma virus. Longer periods of heating delayed the appearance of sensitivity among the imagos. These results are similar to those of LESTRANGE, and BRUN interpreted them in similar fashion. He proposed that the rho virus in cells lends them immunity to superinfection and that heating could destroy rho virus and produce some virus-free cells. These cells would yield high titers of virus when invaded by the infectious sigma variant. The sigma variant arose by recombination or mutation from the rho virus in the flies that spontaneously became CO_2-sensitive. Longer periods of heating, however, destroyed so much rho virus that sigma virus rarely appeared internally to initiate CO_2-sensitivity.

Significantly, BRUN found, as had MERCIER-PAROT (1951), LESTRANGE (1963), and L'HÉRITIER and SIGOT (1946), that treatments applied to developing embryos gave exceptional results. Imagos followed that took longer than controls to manifest CO_2-sensitivity in response to superinfection. Their immunity was apparently greater. BRUN offered this result without satisfactory explanation. The phenomenon could be interpreted as meaning that virus nucleic acid undergoes enhanced multiplication during embryogenesis at elevated temperature. In this event more imaginal cells would be protected by rho-type virus. For sigma infections, a higher virion content would follow in the imagos, as was observed. This interpretation has probably not been made before because dividing cells generally seem to be most easily freed of virus. However, the embryo, especially before cell membrane formation, may be exceptional.

5. Evidence Concerning Virus Transmission

A. Influence of Fly and Virus Mutations

The mutant fly gene *re* limits the hereditary transmission of the virus. That is, flies heterozygous or homozygous for the mutation produce a higher frequency of virus-free eggs than do wild-type controls. (Speculations concerning the mechanisms whereby this gene causes the depletion of germ-line virus were presented in Section 4.)

It is notable that OHANESSIAN-GUILLEMAIN (1963) showed that the rho infection of flies responded like the sigma infection in that the *re* gene caused virus-free eggs to be produced. This evidence supported BRUN's contention that the rho factor is merely a sigma variant. Presumably, if the rho factor were not sigma, the response would have been discernibly different.

The virus genome, too, influences the hereditary transmission of the virus For example, GOLDSTEIN (1949) showed that males inheriting the infection included only virus of a type named v^+ in their sperm; type v^- was excluded. Backcrosses showed that the trait depended on the genotype of the virus, not the fly. GOLDSTEIN showed that v^- virus was capable of invading developing oocytes, but the mechanism for the exclusion of v^- virus from sperm is still unknown.

BUSSEREAU (1964) showed that males infected via their father's sperm contained infectious material in the somatic tissue of their testes although not in their sperm.

Even when the virus is v^+, transmitting males usually do not include it in all of their sperm. The fraction of virus-bearing sperm they produce is called their valence. PLUS showed (Fig. 1) that sperm usually contain only one or few IU. These are actually within the sperm, not merely in the seminal fluid, as L'HÉRITIER (1945) showed from the results of mixed inseminations. SIGOT (1953) reported that valence decreased with age and that heat treatments could accelerate the process. He concluded that virus tended to be lost during the many cell divisions leading to sperm formation.

When sigma is inoculated into females, virus of genotype g^+ will invade the germ line and appear in the progeny (PLUS, 1954, 1955; BRUN, 1963; OHANESSIAN-GUILLEMAIN, 1963). This event has been termed "germinal passage". CO_2-sensitive progeny first appear from eggs laid some time after the formation of IU in the inoculated fly. If a pure strain of g^- virus is inoculated, CO_2-sensitive progeny will not appear. If the inoculated virus is g^+ and also pr^- the frequency of CO_2-sensitive progeny increases for several days, then declines toward zero during the remainder of the flies' lifetime. If the virus were g^+ and pr^+, the frequency of CO_2-sensitive progeny may reach 100 per cent and be maintained there for a period of weeks before falling slightly. The fact that the egg invasion occurs in a peak of activity has been taken to mean that invasive virus is formed only once following inoculation and, therefore, only one crop of virus is ever formed following inoculation. It seems possible, though, that flies stemming from an oogonial infection could sustain virus multiplication throughout their imaginal lives.

These authors suggested that the delay preceding the appearance of CO_2-sensitive progeny means that oocytes are infected not by the inoculum but by newly synthesized virus. They argued that oogonia are not infected because the frequency of sensitive progeny eventually declines. These authors did not measure the fertility or fecundity of the inoculated flies as did JUPIN (personal communication) and so did not comment on such effects. JUPIN showed that the oocyte invasion took place before stage 10 and impaired the fecundity but not the fertility of the females.

The picture of germ-line invasion is further obscured by the fact that g^+ and g^- virus interact when a mixture of the two types is inoculated into females (DUHAMEL and PLUS, 1956; OHANESSIAN-GUILLEMAIN, 1963). The recovery of progeny carrying g^+ virus is reduced, a few progeny are recovered that carry g^-, and none are found with both. The basis for the interaction and the resultant recovery of g^- are unknown.

For this discussion, it is important to emphasize that g^- virus was shown not to invade the germ line by invading oocytes. However, g^- virus presumably can invade germ cells during their formation in the embryo, if present then.

ICONOMIDIS and L'HÉRITIER (1961) used the fact that g^- virus does not invade oocytes to deal with the event whereby virus is lost from the germ cells of females. They wished to know whether the germ cell lost all virus or whether the virus in the germ cell changed into a form which no longer infected germ cells but could still affect somatic cells. If the oogonia of g^+-bearing females should become virus-free, oocytes derived from them could be invaded secondarily by somatic virus. The secondary infection would obscure any new virus formed at the event, termed "décrochage," which led to the loss of the germ-cell infection. Therefore, they examined the progeny of females maintaining a germ-line infection of g^- virus. If décrochage led to the postulated new virus form, it should have been manifested as virus-infected nontransmitting progeny. If décrochage led to virus-free cells, this kind of progeny should be missing.

An examination of the progeny of females maintaining a germ-line infection of g^- virus showed that nearly all of the infected daughters produced some virus-infected eggs. The near absence of virus-infected daughters that produced no virus-infected eggs meant that none of the parental germ cells yielded daughters with infected somatic tissue but not infected germ cells. No virus remained after décrochage which was capable of infecting somatic but not germinal cells. They concluded that the virus maintained by the germ cells did not undergo any transformation at décrochage but was entirely lost from the cell.

Their results were not certain because a small number of infected female progeny were recovered which did not transmit the infection. These may have contained the new virus form and not have been merely females showing 100 per cent décrochage as the authors suggested. However, even if no such new form of virus exists, the experiment does not establish the nature of the event that leads to décrochage, but tells only that the postulated virus is not formed at that time. The origin and loss of the germ line infection remains obscure.

In a separate section of their report, Iconomidis and L'Héritier investigated the frequency of décrochage among females to determine when these events took place. They noted that décrochage taking place in a primary germ cell would lead eventually to many virus-free oogonia and then to a sibship of progeny that included many virus-free flies. Furthermore, any sibship they recovered which included less than five per cent of virus-free flies must have arisen from décrochage that took place late in oogenesis, not earlier than the oogonium stage. Subships containing more than 10 per cent of virus-free flies (early décrochage) were expected to occur about 10 times as frequently as sibship showing late décrochage, if décrochage was equally frequent during the history of the germ line. This estimate followed from considerations of the number of primary germ cells and oogonia in a female.

In fact, the number of sibships indicating early décrochage was found to be greater than 10 times those indicating late décrochage. The authors concluded that virus was most likely to be lost from the germ line early in development, probably at the time of germ cell formation. This may be true, but does not seem to follow necessarily from the data. Late décrochage events might take place often in some females; many of their progeny would be virus-free, and the subships would be classed mistakenly as indicating early décrochage.

Ohanessian-Guillemain (1959, 1963) took advantage of the differences among virus strains to examine infections initiated by a mixture of virus types. She used a virus strain carrying the markers pr^+ p^+ v^- g^+ Ts and another strain carrying pr^- P^- v^+ g^- Tr. The mixed infections were brought about by three different techniques, and a variety of information was obtained.

In one experiment an extract containing the two types of virus was inoculated into females, and the virus types appearing in their progeny were classified. Of 42 infected progeny, 41 carried one or the other parental types. This showed that little or no recombination took place between the virus types while multiplying in the females. It showed, too, that germ cells invaded by virus from inoculated females usually carry clones of a single virus type. Either a germ cell is invaded by only a single particle, or only one type of particle of those invading succeeds in multiplying.

One of the 42 infected progeny proved to contain a mixture of the parental virus types. Since no new types appeared during its development, the result suggests that recombination is not frequent following germ line invasion by two virus types. This conclusion, of course, is tenuous. However, this one fly illustrated that germ-line invasion does not always yield virus clones in the progeny.

The third technique used to generate a mixed infection was to inoculate females with one virus type, then cross them to males transmitting the other. Examination of clones from doubly infected progeny showed that they contained very little (1/68) of the paternal type, presumably because it entered the egg after the maternal contribution was already multiplying. Seven of the clones, representing at least four different events, were found to carry nonparental virus types. These new types were too numerous to have been derived from mutations or from minority virus populations in the original extracts and, therefore, were considered recombinants. The new types were clonally distributed in the offspring, suggesting that recombination takes place early in embryology. However, the new virus types recovered were not completely stable, and each gave rise to segregants for one of the three markers followed. To account for this, OHANESSIAN-GUILLEMAIN proposed that these virus types were heterozygous for the segregating factors (partial diploids).

From a review of these mixed infections, it is clear that the data were too few to allow the conclusions to be certain. The experiments were done *in vivo* during the course of multiple virus and fly generations, with little understanding of the events taking place at the cellular level. However, these interprepretations can stand until new information challenges them.

BRUN (1963) also reported studying the hereditary transmission of mixed infections. He found that several strains of rho-infected flies carried in the laboratory maintained a population of virus of types mixed for the plaque size character. A spectrum of these types could be maintained for years within the germ lines of each of the flies. When a fly line homogeneous for a plaque type or for a new character like ultra-rho appeared, however, it seemed to do so in a single step rather than by a gradual reduction in the number of virus types.

The maintenance of this variability at an equilibrium is difficult to explain because genetic drift would tent to create homogeneous virus populations when the virus particles were distributed among germ cells in each generation. Drift could not be overcome by mutation toward heterogeneity because mutation rates equal to the reciprocal of the number of particles would be required. Such rates would be enormous, perhaps greater than one every 100 particle duplications, as BRUN calculated only several dozen particles per oogonial cell. BRUN'S calculations were based on the assumption that virus particles duplicated and were distributed at each of 15 cell generations during each Drosophila generation. He concluded that even if some mechanism insured that half of the virus particles duplicated and were divided equationally at each division, the mutation rate required to maintain heterogeneity between fly generations was too great to be possible.

BRUN rejected the hypothesis that virus is transmitted by units containing several virus genomes, because changes in virus behavior appeared in single steps as if they were effected by a single "mutation". Instead, he favored a hypothesis whereby each virus particle could be a partial diploid and thereby carry a multiplicity of information. The changes in virus behavior that appeared, from rho to ultra-rho, for example, could then arise by recombination during virus multiplication.

BRUN'S model of partial diploidy best explained his results concerning the maintenance of heterogeneity in some lines of infected flies and, interestingly, OHANESSIAN-GUILLEMAIN'S study of mixed infections led to a similar conclusion. However, BRUN'S calculations did not dispose of the possibilities that a heterogeneous population of haploid genomes could be maintained by mutation or by some special mechanism.

BRUN pointed out that the values he used for calculations are only estimates. For example, Drosophila germ cells are kept isolated during the divisions that take place in the fly somatic tissues during development. BRUN'S estimate of several dozen particles may, therefore, have been subjected to only three or four pole cell divisions in each fly generation (DEMEREC, 1950) and have had few chances to drift. This would be especially important if the females laid few eggs in each generation, as JUPIN'S results suggest might occur. Furthermore, the estimate of several dozen particles per oogonial cell may represent the lowest number to which the virus population falls during the history of the germ line. That is, the oogonia may receive their virus population from a much larger one in the egg cytoplasm during germ cell formation and, therefore, drift would be important only in the oogonial divisions.

On the other hand, the virus population in the oogonia is probably so small that mutation could not maintain heterogeneity, especially over years. In this case virus might be transmitted by units that each contained several virus genomes. For example, virus might be attached to cell organelles. BRUN'S single-step changes could be explained as exceptions to this rule and due to the inclusion in an oocyte of one or few genomes of a given type. In fact, many other models could be invented to account for specific inclusion of virus genomes into daughter cells, but none are testable yet.

Using different virus strains, VIGIER (1966) followed the transmission of infections mixed for plaque types through numerous fly generations. He found that a clone of virus of intermediate plaque type gave rise to pure lines of larger and smaller types at a frequency such that the half life of the intermediate type line was about four generations. Assuming that the population in the germ line was divided at random during 15 cell divisions in each fly generation, he calculated that each germ cell contained an average of 30 virus particles, in good agreement with BRUN'S figures.

However, flies containing the mixed infection were not eventually eliminated, but were finally maintained at an equilibrium value. VIGIER calculated that mutation rates on the order of one in every 1,000 particle duplications would be required in order to maintain the observed equilibrium. He believed that gene mutation could not be this frequent, and therefore some additional mechanism must be invoked to explain the equilibrium. He could not adopt BRUN'S hypothesis of recombination in a duplicated segment because some of his lines drifted reversibly, never becoming pure stable plaque types as lines made homozygous by recombination presumably should be. VIGIER could explain his results by attributing them to phase changes such as take place in bacteria, but decided that the stability of some of the virus characters BRUN studied precluded this as a general explanation. His demonstration that sigma virus may remain unstable, apparently for an unlimited number of duplications, therefore remained unaccountable.

He showed, too, that the virus phenotype was highly variable during its multiplication in somatic cells, changing at rates up to one mutation per 100 particle duplications. No good explanation can be proposed at present except, perhaps, to observe again that variability might result from virus genomes being transmitted as groups. Since VIGIER cloned his virus by germ-line passage and by dilution this would mean that his techniques were not successful. If this were so, the lines he presumed were pure were indeed mixed, and the variability observed was merely recombination and assortment.

The dilution technique would often fail to yield clones if virus genomes were physically associated into groups in homogenates. There is reason to think that such groups do exist as both PLUS (1960) and I (1962) have reported the infectivity to be associated with a particulate fraction in homogenates. The cloning technique by germinal passage is more difficult to discount. However, germinal passage is not well understood. Synergistic effects between viruses may be involved in the process (g^+, g^-) and, as detailed earlier, OHANESSIAN-GUILLEMAIN found that germinal passage sometimes resulted in mixed infections. If the cloning techniques were not faulty and the measurements of plaque size were reliable, VIGIER'S results indicate, obviously, that the sigma virus phenotype can be enormously variable.

Another investigation of mixed infections was reported by me in 1966. I followed a strain of flies which remained mixed for plaque-size virus types for 20 generations, during which time the proportions of plaque types showed little change. Fast growers made up a majority of the population (about 87 per cent) and the females included virus in nearly all of their eggs. At the same time, two infected strains of a different fly genotype were selected for the property of producing only virus-infected eggs. Compared to the mixed strain, one of these was enriched for the fast-growing virus type and the other for the slow-growing type. Despite the selection practiced, females from the two relatively pure strains continued to produce more virus-free eggs than those from the mixed strain. Since neither enrichment for fast growers or slow growers led to females that produced only virus-infected eggs, I concluded that plaque size alone was not sufficient to insure continued oogonial infection. Either the fly genotype made a contribution to such infection, or some necessary special event or virus type did not occur in the two relatively pure lines during the experiment.

B. Influence of Temperature

BRUN (1962) examined the effects of temperature on the transmission of the sigma infection by females. He crossed females that inherited g^- virus to virus-free males, subjected them to 30° C for periods up to 16 days, then returned them to 20° C to lay eggs. The first eggs laid at 20° C had been nearly mature when heated; eggs laid later had been developing oocytes when heated, and eggs laid after about 12 days represented those heated as oogonia. CO_2-resistant progeny, representing cured germ cells, were recovered and their distribution examined to discover the thermosensitivity of the infection at each stage of oogenesis.

On the first day after return to 20° C, eggs laid by females transmitting "weak" virus gave rise to many CO_2-resistant, presumably virus -free progeny. CO_2-resistant progeny (virus-free eggs) became less frequent in the next few days until all progeny appearing were CO_2-sensitive. Then, CO_2-resistants reappeared and their frequency increased to a plateau level. This level represented a frequency of oogonial cell cures and was proportional to the duration of the treatment. Females transmitting "strong" virus gave similar germ-cell cures except that no CO_2-resistants were observed in the eggs laid just after the return to 20° C.

BRUN argued then that each oogonial cell contained more than one unit of virus, that these multiplied during oogenesis, and that heating acted to cause a progressive loss in virus during the treatment. This hypothesis accounted for the virus-free eggs laid just after the return to 20° C; their virus content had been reduced to zero by the heating during oogenesis. Virus-infected eggs then followed because these had been subjected to heat during the treatment for a shorter time during oogenesis, a time not long enough to reduce their content to zero. Finally, virus-free progeny appeared from oogonia that were cured before their content of virus could multiply during oogenesis.

When the progeny were followed for another generation, BRUN found that all of the virus-infected females continued to transmit virus to their progeny with no increase in CO_2-resistants. Some exceptional females were found, however, among those from eggs laid immediately after the return to 20° C by treated females carrying weak virus. These exceptions threw an increased number of virus-free offspring. BRUN proposed that the exceptional females contained so few virus units during the time their germ cells were formed that some of their germ cells received no virus. Therefore, they became germinal mosaics and produced mixed progeny. BRUN suggested they were somatic mosaics too, as embryos, but this was obscured by the infective process in the soma.

Finally, BRUN pointed out that a hypothesis whereby oogonia each carried only one virus unit did not readily account for the appearances of cured progeny from heated germ cells. Virus thermosensitivity would have to be assigned arbitrarily to the stages of oogonium and mature egg and this could not be justified without making more *ad hoc* assumptions.

BRUN plotted the rate at which the heat cures were effected by the duration of the treatments. He obtained typical "multihit" curves showing CO_2-sensitive progeny disappearing slowly at first, then exponentially, as the duration increased. This analysis suggested that the heat treatments caused the progressive loss of independent units of virus in the oogonia, the initial number of units being 16 to 40 per cell.

The data, however, were not precise enough to rule out the possibility that each oogonial cell contained only one unit of virus. A large measure of variability was introduced by the heterogeneity among females, and the methods of calculation were unsatisfactory insofar as they may have oversimplified the complexity of the *in vivo* system. Furthermore, analysis of oogonial cure at 20° C (spontaneous appearance of CO_2-resistants) led to a lower estimate of about five units of virus per cell. Brun concluded that the shortcomings of the experiments did not provide any argument in favor of a single unit of virus per oogonial cell; rather, all the evidence suggested that several were present. It should be noted that Brun's estimates were made on strains of infected flies that were selected because their oogonia could be readily heat-cured. Other fly strains might contain a different number of heat-curable virus units.

Brun (1963) also studied the effects of temperature on the transmission of the rho- and ultra-rho infections. The protocol was similar to that used to study sigma except that, as all of the infected progeny were CO_2-resistant, heat-cured oogonia were detected as progeny infectable by inoculation rather than as CO_2-resistants. Germ cell cures similar to those described for flies infected with a weak strain of sigma virus were obtained. The data led to a best estimate of about 50 virus units per oogonial cell, subject to the same reservations as the estimates for sigma-infected oogonia. Regardless of the absolute number of virus units per cell, however, the transmission of the sigma, rho, and ultra-rho infections were shown to respond similarly, both qualitatively and quantitatively, to heat treatment. This provides a strong argument that all three types of infection are due to variants of the sigma virus, and the germ line contents of the flies carrying these infections differ only by the genetic makeup of the virus. That is, no novel virus-host relationship is likely to account for the differences among these strains because the transmissions of all of them are equally sensitive to heat treatment.

6. Hypotheses Concerning the Infection

A. Synopsis of the Problem

The following passage is a brief summary of some of the experimental facts concerning the sigma infection: Drosophila may somehow become infected with sigma virus, which multiplies in the somatic tissues to yield a final plateau of extractable infectious material. In association with the appearance of infectious material, the flies become CO_2-sensitive and resistant to the multiplication of more sigma virus, should any be inoculated. The virus may be present in the germ line, as evidenced by the hereditary transmission of the infection by the flies. The germ line infection may be accomplished by virus invading either the oocytes or the oogonia of the flies. Once the oogonial invasion is accomplished, oocyte invasion is not necessary to insure the transmission of the infection through generations of flies.

These facts alone would not make the sigma virus infection extraordinary. The phenomena resemble those encountered in other animal virus infections and could be explained by hypotheses which stipulated only commonplace virus behavior. However, several aspects of the infection demand special explanation.

The first is the fact, illustrated in Fig. 1, that oogonial infection is followed by only limited production of extractable infectious material during fly development. The infectious material accumulates slower and to lower final values than it does in flies infected at a later stage of development.

Secondly, the hereditary transmission of the infection is not related simply to the content of infectious material in the flies. The (low titer) flies from an oogonial infection transmit the infection to more of their progeny than do the flies infected at a later stage of development. This shows that high titers during larval and imaginal stages do not insure transmission. An additional factor must be responsible for the maintenance of the oogonial infection through fly generations.

A third unusual aspect of the infection is the existence of ultra-rho flies, which maintain the virus infection for generations in the absence of any extractable infectious material. This pehnomenon is similar to the previous one in the sense that germ cell infection is maintained despite low titer. In addition, it is an outstanding example of a latent virus infection.

There are, too, scattered through the observations that have been reviewed, numerous references to Drosophila resistance to virus multiplication. Flies maintaining an oogonial infection were shown to support virus multplication less well than flies infected via their father's sperm. Rho or ultra-rho-bearing flies are resistant to superinfection by sigma. Imagos maintaining an oogonial infection are immune to superinfection by additional sigma. Flies freed of infectious material by heat treatment are resistant to superinfection. Virus types limit each other's multiplication following a mixed infection. Elevated temperatures may prevent infection. Age lends resistance to imagos, and Drosophila genes may confer resistance too.

The origins and interrelationships of these various resistances are not yet understood. Some may depend on the presence of intracellular virus material and some be responses by the fly which do not require the continued presence of virus for their expression. They must be dealt with by any models that explain the cause of the infection.

B. Hypotheses and Models

Early investigations of the sigma virus infection revealed that infected fly strains differed by the frequency with which CO_2-resistant flies appeared in them. Fly lines in which few CO_2-resistants appeared were termed more stable than those in which many appeared (L'HÉRITIER and TEISSIER, 1945). This line of research has never been followed vigorously, probably because of the difficulties in standardizing mass culture conditions. In general, strains that are stable in this sense seem to be somehow balanced; if the fly genotype is changed by outcrossing, or the virus type by selection, the derivative strain will be less stable (L'HÉRITIER and TEISSIER, 1954; SEECOF, unpublished results).

In mass culture such stability would perhaps be favored by innocuous virus types. Evidence suggests that the virus content of flies in mass culture tends to drift toward slow-multiplying types (VIGIER, 1966). However, flies from mass cultures that yield few CO_2-resistants may contain a higher proportion of fast-growing virus types than strains which yield more CO_2-resistants (SEECOF, 1966). The balance is probably dependent on complex relationships among virus types and fly genomes. It seems likely too, since females that give rise to strains yielding no CO_2-resistants are generally rare, that the necessary kind of germ line infection requires some special event for its initiation.

After initial observations were made on strain stability, experiments were focused on the infection in single flies and their offspring. BRUN and SIGOT (1955) studied pedigrees of flies taken from a strain that produced few or no CO_2-resistants. They found that individual flies could produce all virus-infected eggs during their lifetimes, and this character was true of their descendents too. It was a stable property, characterized further by the presence of relatively low imaginal titers and by transmitting males among the progeny. In contrast, females whose infection originated via their father's sperm could transmit it to only part of their progeny and males infected via sperm could not transmit it further. The sperm-initiated infections also led to higher imaginal titers than did the infection inherited through females. BRUN and SIGOT proposed to call the infectious state characterized by low titers, transmitting males, and maternal inheritance the stabilized state, and the sperm-initiated infection the non-stabilized state.

The germ lines of the nonstabilized females were clearly mosaic in that the females produced both virus-infected and virus-free progeny. Some of their infected daughters also displayed relatively high titers and nontransmitting sons and some of the infected sons failed to transmit. These were nonstabilized flies. However, some stabilized males and females were detected among the progeny too. These arose as unique events in their maternal germ lines and were termed neostabilized flies. BRUN and SIGOT could not determine the origin of the neostabilization but showed that it did not follow from a virus type that was especially efficient at invading developing oocytes. The virus content of stabilized and nonstabilized flies seemed to have equal ability in this respect.

Virus was obviously present in the germ line of stabilized flies because they produced infected progeny. Occasionally, stabilized females produced daughters that behaved in nonstabilized fashion. That is, they transmitted virus to only part of their progeny, their sons were nontransmitters, and their titers were probably high. It seemed possible that these females maintained a virus which infected the somatic tissue but not the oogonia or primary germ cells.

However, ICONOMIDIS and L'HÉRITIER (1961) showed that they probably contained virus like that of their mothers and not a virus type with such unusual properties. Therefore, the nonstabilized flies that arose from stabilized females were considered to maintain the infection in the same nonstabilized state as did the flies that derived their infection from paternal sperm. ICONOMIDIS and L'HÉRITIER also considered that their failure to find such a virus type meant that the nonstabilized flies probably arose from germ cells entirely freed of virus. Their infection was initiated only secondarily by maternal virus invading the oocytes from which they originated.

These conclusions resolved the difference between the stabilized and nonstabilized states into a difference dependent on the source of the infection. Flies inheriting virus from a maternal oogonial infection were stabilized, and flies with infections that originated in infected oocytes or via paternal sperm were nonstabilized.

Now, by definition, stabilized flies are those maintaining an oogonial infection regardless of the number of virus-free gametes they produce. Furthermore, BRUN (1963) argued that the rho and sigma infections could be maintained in the germ line in a noninfectious state and the stabilized condition was, therefore, a virus-cell relationship maintained for cell generations without benefit of an infectious process.

The term stabilized has obviously evolved to a point in which it is charged with meaning. However, the concept of the stabilized condition still fails to account for many of the critical aspects of the sigma virus infection. We still need to speculate on the nature of the events which lead to the initiation or loss of the germ line infection, on the resistance to virus infection, and on the identity of the infectious and noninfectious forms of the virus.

A comprehensive model for the sigma virus infection was proposed by BRUN (1963) who suggested that virus could be maintained in a cellular association he called the stabilized state. Flies could maintain virus in this condition, in their germ lines, for generations without the intervention of any infectious process. Ultra-rho-infected flies do behave in just this way, transmitting the infection in the absence of any content of extractable infectious material. BRUN could free ultra-rho-infected germ cells with heat treatments and calculated from the rate of „cure" that each germ cell contained about 40 independent virus units.

The virus responsible for the hereditary transmission of the sigma infection, contained within their germ cells, responded to heat treatment in a fashion identical to that of the virus in the ultra-rho-infected germ cells. BRUN concluded that the sigma variant was transmitted in the same stabilized state as the ultra-rho infection. Since some germ cells of a fly transmitting the virus in stabilized conditions could be virus -free, the stabilized state was not at the level of the whole fly but on a cellular level.

He postulated that most of the somatic cells of a fly transmitting the stabilized virus condition probably maintained the stabilized infection as well, having obtained it during cell formation in the embryo. Although the virus material stabilized in these cells did not participate in the germinal transmission of the virus, he ventured that it lent somatic cells their observed immunity to superinfection. Presumably the stabilized particles inhibited, at least partially, the elaboration of infectious material by the cell.

BRUN accounted for the appearance of infectious material during the development of sigma-infected stabilized flies by suggesting that it was formed from the stabilized particles within cells in a series of inductions. The infectious virus cells thus formed could then multiply in a virus-free cell or perhaps initiate the limited production of infectious material in a stabilized cell.

BRUN'S conclusions concerning the stabilized state do not depend for their validity upon the details of its establishment or loss. His model is particularly convincing when presented as integrated with the events in Drosophila embryology, especially those of germ cell formation as suggested by SIGOT (1953). BRUN'S model accounts for the enhanced transmission of the infection by flies originating from oogonial infection in the following way: Only embryos derived from infected oogonia would have sufficient virus within them at the time cell membranes were formed to insure that virus was included in most cells. Virus introduced by germ-cell passage or sperm would not usually have multiplied sufficiently by the time of cell formation to be included in germ cells. This would mean that the germ line infection is re-established in every fly generation by the inclusion of virus in the primary germ cells.

Coincidentally, at the time of cell membrane formation, stabilized virus would be included in many somatic cells. These particles impede free virus multiplication during fly development, accounting for the relatively slow accumulation of infectious material in flies derived from infected oogonia. The model would account too for the continuance of the ultra-rho- infection in the absence of infectious material. Flies with this infection would simply contain a form of the virus that was noninfectious in the assay procedure but capable of multiplying intracellularly.

When presented in this way, BRUN'S conclusion that the stabilized state is maintained in the absence of infection is open to question. The stabilized state might need to be re-established in each embryo by an invasion of the germ cells by some form of virus. Alternatively, virus might be passively included in the primary germ cells when their cell membranes were formed. In the latter event, the decision as to whether they have been "infected" becomes only a semantic difficulty. Indeed, both events may occur. The important consideration here is that the events during embryology are critical to understanding the hereditary transmission of the virus and deserve close study in the future.

Fig. 2 presents this model diagrammatically. The cycle at the left shows flies maintaining an oogonial infection. They contain noninfectious virus (v) in all of their germ cells (shaded). During cell formation in the embryo, most somatic cells receive virus, and infectious material (σ) is formed during development. The cycle at the right shows the infection initiated by oocyte invasion or via sperm or inoculation. The flies maintain σ in their somatic cells, but their germ cells remain virus-free unless invaded as oocytes. If virus is present in the embryo early enough, the germ line infection may be initiated. The infected germ cell may then originate a line of oogonially-infected flies.

The proportion of infected offspring may reflect merely the quantity of virus present in the embryo during germ cell formation. This possibility was favored by BREGLIANO (1965) who found that the oocytes of flies transmitting an oogonial infection contain relatively high titers compared to those infected by invading virus from an inoculum. On the other hand, it seems unlikely that the efficiency of transmission depends on virus quantity alone, because lines of flies in which all individuals inherit virus are sometimes difficult to establish, even by deliberate selection (SEECOF, 1966). Furthermore, OZOLINS showed that *re/re* flies transmitted their oogonial infection as efficiently as did their *re/re*$^+$ mothers despite a lower titer in their oocytes. These facts suggest that additional factors may operate to insure that the primary germ cells receive virus.

If some special mechanism is thought to exist to insure the infection of primary germ cells, the door is open to speculations concerning a possible role

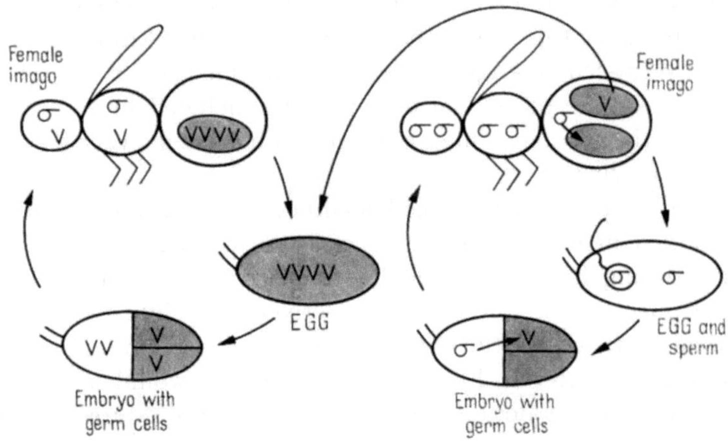

Fig. 2. Hereditary transmission of the sigma virus. Shaded cells are germ cells. V means noninfectious virus, possibly "integrated" somehow in the cell. σ means infectious virus, probably virions

for an integrated virus. Such speculation immediately suggests an analogy to bacterial lysogination and a chromosome site for sigma. So far, however, it has not been possible to show such a site for sigma. Virus which might be attached to paternal chromosomes cannot be followed in crosses because sperm-born virus never appears in the gametes of the following generation except as a secondary infection. If virus is attached to paternal chromosomes in sperm, it must be lost from them after fertilization. Virus which might be attached to maternal chromosomes cannot be followed because paternal chromosomes introduced into an infected egg presumably could be infected by the high egg virus content. This would obscure the linkage data.

Even if no chromosomal site exists for sigma, the primary germ cells might be better assured of infection if virus were associated with a cell organelle. For example, if some virus particles succeeded in associating with mitochondria they might be better assured of distribution to daughter cells and better protected from degradative intracellular enzymes. If virus particles were sometimes associated with some organelle, the transfer of the unit could account for the maintenance of genetic variability despite drift. Even cellular resistance to superinfection might follow from the organelle sites being occupied and thus unavailable to the invading virus.

Another model, related to BRUN'S, has been advanced to account for the phenomena associated with the virus infection. VIGIER (1961, 1966) considered that his results showed BRUN'S hypothesis of successive inductions inadequate to account for the entire somatic content of infectious material. He decided that at least part of the infectious material was derived from a process of intercellular infection. LESTRANGE (1963) found that the infectious material in flies stemming from an oogonial infection responded differently to heat treatment than that in flies infected via sperm. A model like BRUN'S was not sufficiently elaborate to explain these responses. L'HÉRITIER (1962) attempted to encompass the results of VIGIER and LESTRANGE with a model that assigned special properties to the germ line virus.

L'HÉRITIER proposed that the noninfectious virus component responsible for the hereditary transmission of the stabilized state (called the genetic virus) was probably confined to the germ line. Ordinary cycles of virus multiplication yielded virions in the somatic tissue, but these did not occur in the germ cells. The genetic virus was responsible for infecting the somatic tissues, and ordinary cycles of virus multiplication yielded virions in them. The multiplication of genetic virus, however, did not lead to virion production within the germ cells or within the posterior part of the oocyte that was destined to be germ cells in the embryo.

This model accepts the same concept of the stabilized state as does BRUN'S and is compatible with the scheme illustrated in Fig. 2. It cannot account for all of LESTRANGE'S unusual findings, but does provide an avenue for somatic virus to multiply independently of the germ line virus. The model is more complicated because it assigns special properties to the germ line virus, and requires a humoral factor to account for the limitation on virus multiplication in stabilized flies.

A humoral factor, if one exists, might be demonstrable. For example, an extract from flies maintaining an oogonial infection could be mixed in various proportions with an extract from flies infected via sperm. A humoral resistance factor in the former extract might lower the apparent titers of the mixtures more than could be accounted for by dilution.

The various speculations considered in this section provide a coherent picture of the infection. The efficient virus transmission by females maintaining an oogonial infection, despite their low imaginal titers, is seen as a consequence of a sustained germ line infection. The factors which insure this infection are still in doubt. The low imaginal titers of these flies seem to follow from the germ line infection. The germ line virus provides resistance to virus multiplication either by a humoral factor or by its presence in the somatic cells, but no details are known. Many other individual phenomena concerned with the infection still lack explanation, but when details of the intracellular behavior of the virus become known, hopefully they can be fitted into this picture.

One aspect deserving further comment is the manifestation of the noninfectious form of the virus, most clearly demonstrated in rho-infected flies, where the entire virus content is noninfectious. Its presence in other flies can only be surmised. All infected flies might contain a component of noninfectious virus in addition to their infectious material. No way yet exists to measure the total content of the noninfectious material in flies. Possibly flies with low titers contain large amounts of it, those with high titers relatively less; and all infected flies, therefore, contain about the same number of virus genomes. If so, any resistance to virus multiplication might be just resistance to virus maturation.

Then there is the question as to whether the noninfectious form is an immature or incomplete virus which cannot initiate a complete cycle of virus multiplication, or whether it is a form which competes with the sigma variant as a parallel infection. The alternate virus form would be one which never rendered flies CO_2-sensitive, perhaps because it failed to reach a target or did not provoke the biochemical reactions resulting in CO_2-sensitivity. Such a form would be invisible in the assay, nondetectable rather than noninfectious. It should be mentioned that Brun (1963), when considering plaque-type variants, could rule out any form that was merely less efficient in causing CO_2-sensitivity. If the noninfectious form is really a parallel infection by an infectious sigma variant, competing in both transmitting and inoculated flies, then the concept of the stabilized state must be discarded.

All efforts made so far to detect a competing form which did not confer CO_2-sensitivity have given negative results. None were detected in an extract taken from flies with a weak plaque-type infection where it should have been prominent (Seecof, 1962). This study showed that it would have had to fail to confer immunity to superinfection as well as fail to render flies CO_2-sensitive. In a different study, I showed that it also failed to interfere with fly development (1966). It seems unreasonable to assign three special properties to a competing form, so the interpretation that the nondetectable form is a parallel infection is probably wrong.

This conclusion returns us to a consideration of the identity of the noninfectious form. One guess would still be that it was virus nucleic acid or some other incomplete particle. Alternatively, it might be that incomplete particles or even virions are noninfectious in the assay procedure if they become associated with a cellular fraction or organelle. An experimental confirmation of any of these possibilities would provide important insights into latent virus infections.

C. Prospects for Future Investigations

Nearly all evidence gathered so far concerning the sigma virus infection has been based on measurements concerning the CO_2-sensitivity of flies. This has led to a description of virus production and hereditary transmission at the level of whole flies. We now need to know the details of virus multiplication on a cellular level in order to interpret the phenomena yet unexplained and to unify the facts at hand.

Until recently all information was gathered without the benefit of powerful techniques that are used routinely for studying other viruses. These kinds of experiments are now within reach and prospects look good. BERKALOFF et al. (1965) have already made electron micrographs of the virus, and future photographs should lead to the determination of particle-to-infectivity ratios. When the assay for sigma is based on virions instead of infectious units, all titer measurements should be unambiguous.

OHANESSIAN and ECHALIER (1967) recently claimed to have observed sigma virus multiplication in cell cultures. When details of their procedures are published, rapid advances can be expected. Virus multiplication *in vitro* will lead to pure preparations of virus and to detection of the virus via isotope labeling and fluorescent antibody techniques. The identity and distribution of sigma nucleic acid should become apparent, and the description of the virus life cycle will be established. A different technique with similar promise was reported by BERNARD (1964), who showed that virus would multiply in imaginal discs when these were kept as implants in imagos.

Research on the sigma virus infection is also being pursued in an entirely new direction. Morphologically related viruses are being studied to determine if they can be maintained by Drosophila. In one such experiment, PÉRIÈS et al. (1966) reported that vesicular stomatitis could multiply in the flies but they did not become CO_2-sensitive.

PLUS and ATANASIU (1966) inoculated Drosophila with rabies virus and found that the flies' fertility, fecundity, and longevity were reduced, although they did not become CO_2-sensitive. These effects were caused by a virus which multiplied within the flies and was transmitted to the offspring. The virus was probably rabies virus, but certain identification was not possible because it was no longer pathogenic to mice after passage in flies. PLUS and ATANASIU noted that the results suggested an insect reservoir for rabies virus, and indicated that this possibility would be pursued.

Studies of flies infected with both sigma virus and another virus may lead to an especially interesting result. Prior infection by sigma virus may lend resistance to the multplication of the second virus. Such a protective effect could explain why the sigma virus infection is so widespread under natural conditions despite the deleterious effects it produces in flies. Perhaps a latent sigma virus infection protects the animal against infection by many other more virulent viruses.

References

BERKALOFF, A., J. C. BREGLIANO, et A. OHANESSIAN: Mise en évidence de virions dans des Drosophiles infectées par le virus héréditaire σ. C.R. Acad. Sci. (Paris) **260**, 5956—5959 (1965).

BERNARD, J.: Propagation du virus héréditaire de la Drosophile, au niveau de disques imaginaux transplantés. R. C. Acad. Sci. (Paris) **259**, 4879—4881 (1964).

BREGLIANO, J. C.: Contribution à l'étude du virus héréditaire de la Drosophile. Cinétique de l'accumulation du virus infectieux dans une mouche stabilisée. C.R. Acad. Sci. (Paris) **256**, 2942—2944 (1963).

RREGLIANO, J. C.: Étude de la quantité de virus infectieux contenue dans la cystes ovariens de Drosophiles stabilisées et inoculées. Ann. Inst. Pasteur **109**, 638—651 (1965).

BRUN, G.: Étude d'une association du virus et de son hôte la Drosophile: l'état stabilisée. These Biol. exp., Orsay (S.-et-O.) (1963).

—, et A. SIGOT: Étude de la sensibilité héréditaire an gaz carbonique chez la Drosophile, II. Ann. Inst. Pasteur **88**, 488—512 (1955).

BUSSEREAU, F.: Essai d'extraction de virus infectieux à partir des spermatozoïdes de Drosophiles mâles stabilisées. C.R. Acad. Sci. (Paris) **259**, 3888—3891 (1964).

DEMEREC, M.: Biology of Drosophila. New York: Wiley Publ. Co. 1950.

DUHAMEL, C.: Étude de la sensibilité héréditaire à l'anhydride carbonique chez la Drosophile. Description de quelques variants du virus. C. R. Acad. Sci. (Paris) **239**, 1157—1159 (1954).

—, et N. PLUS: Phénomène d'interférence entre deux variants du virus de la Drosophile. C. R. Acad. Sci. (Paris) **242**, 1540—1543 (1956).

GOLDSTEIN, L.: Contribution à l'étude de la sensibilité hereditaire au gaz carbonique chez la Drosophile. Mise en evidence d'une forme nouvelle du genoïde. Bull. biol. France Belg. **83**, 177—188 (1949).

GUILLEMAIN, A.: Découverte et localization d'une gène empêchant le multiplication du virus de la sensibilité héréditaire au CO^2 chez D.m. C. R. Acad. Sci. (Paris) **236**, 1085—1086 (1953).

HACKETT, A. J.: A possible morphologic basis for the autointerference phenomenon in vesicular stomatitis virus. Virology **24**, 51—59 (1964).

HOWATSON, A. F., and G. F. WHITMORE: The development and structure of vesicular stomatitis virus. Virology **16**, 466—478 (1962).

HUGON DE SCOEUX, F.: Contribution à l'étude de la sensibilité héréditaire au gaz carbonique chez la Drosophile. Essai de purification du genoïde. Rev. canad. Biol. **11**, 1—13 (1952).

ICONOMIDIS, J., et PH. L'HÉRITIER: Les relations du virus σ avec son hôte Drosophile. Étude de la perte spontanée de l'état stabilisé. Ann. Génét. **2**, 53—58 (1961).

KALMUS, H., J. KERRIDGE, and F. TATTESFIELD: Occurrence of susceptibility to carbon dioxide in D.m. from different countries. Nature (Lond.) **173**, 1101—1102 (1954).

KING, R. C., A. C. RUBINSON, and R. F. SMITH: Oogenesis in adult *Drosophila melanogaster*. Growth **20**, 121—157 (1956).

LESTRANGE, M. T. DE: Action de la température sur le virus responsable de la sensibilité à l'anhydride carbonique chez la Drosophile. C. R. Acad. Sci. (Paris) **239**, 1159—1162 (1954).

— Contribution à l'étude de virus héréditaire de la Drosophile: Action de l'hyperthermie sur le contenu en virus des tissus somatiques de l'hôte. Ann. Génét. **6**, 39—96 (1963).

L'HÉRITIER, PH.: Les relations du virus héréditaire de la Drosophile avec son hôte. Ann. Inst. Pasteur **102**, 511—526 (1962).

—, et F. HUGON DE SCOEUX: Transmission of the carbon dioxide sensitivity of Drosophila by grafting. Nature (Lond.) **157**, 729 (1946).

— — Transmission par greffe et injection de la sensibilité héréditaire au gaz carbonique chez la Drosophile. Bull biol. France Belg. **81**, 70—91 (1947).

—, et N. PLUS: Inactivation par les rayons X du virus responsable de la sensibilité au CO^2 chez la Drosophile. C. R. Acad. Sci. (Paris) **231**, 192—194 (1950).

—, et A. SIGOT: Contribution à l'étude du phénomène de la sensibilité au CO^2 chez la Drosophile. Influence de chauffage aux différents stades du développement sur la manifestation de la sensibilité chez l'imago. Bull. biol. France Belg. **80**, 172—227 (1946).

L'HÉRITIER, PH., et G. TEISSIER: Transmission héréditaire de la sensibilité au gaz carbonique chez *Drosphila melanogaster*. Publs. lab. Ecole Normale supériore. Paris, Biol., **1**, 35—74 (1945).
MERCIER-PAROT, L.: Action de la température sur la transmission par les Drosophiles femelles de la sensibilité héréditaire au gaz carbonique. Bull. biol. France Belg. **85**, 226—236 (1951).
OHANESSIAN, A., and G. ECHALIER: Multiplication of Drosophila hereditary virus (σ virus) in Drosophila embryonic cells cultivated *in vitro*. Nature (Lond.) **212**, 1049—1050 (1967).
OHANESSIAN-GUILLEMAIN, A.: Sensibilité héréditaire à l'anhydride carbonique chez la Drosophile. Étude d'une mutation du virus σ modifiant son spectre d'activité. R. C. Acad. Sci. (Paris) **243**, 1922—1924 (1956).
— Étude génétique du virus héréditaire de la Drosophile (σ) mutations et recomcombinaison génétique. Ann. Génét. **1**, 59—68 (1959).
— Étude de facteurs génétiques controlant les relations du virus σ et la Drosophile son hôte. Ann. Génét. **5**, 1—64 (1963).
PÉRIÈS, J., P. PRINTZ, M. CANIVET et J. C. CHUAT: Multiplication du virus de la stomatite vesiculaire chez *Drosophila melanogaster*. C. R. Acad. Sci. (Paris) **262**, 2106—2107 (1966).
PLUS, N.: Courbes de multiplication de la sensibilité au CO^2 chez la Drosophile. C. R. Acad. Sci. (Paris) **233**, 1489—1491 (1951).
— Étude de la multiplication du virus de la sensibilité au gaz carbonique chez la Drosophile. Bull. biol. France Belg. **88**, 248—293 (1954).
— Étude de la sensibilité héréditaire au CO^2 chez la Drosophile, I. Ann. Inst. Pasteur **88**, 347—363 (1955).
— Utilization des méthodes de séparation des organites cellulaires pour la purification du virus σ de la Drosphile. C. R. Acad. Sci. (Paris) **251**, 1685—1686 (1960).
— Comportement du virus σ de la Drosophile en centrifugation à gradient de densité de saccharose. C. R. Acad. Sci. (Paris) **255**, 3498—3500 (1962).
— Action de la 5-fluorodesoxyuridine sur la multiplication du virus σ de la Drosophile. Biochim biophy. Acta (Amst.) **72**, 92—105 (1963).
—, et P. ATANASIU: Sélection d'un mutant du virus rabique adapté à un insecte: *Drosophila melanogaster*. C. R. Acad. Sci. (Paris) **263**, 89—92 (1966).
SEECOF, R. L.: CO_2 sensitivity in Drosophila as a latent virus infection. Cold Spr. Harb. Symp. quant. Biol. **16**, 99—111 (1962).
— Deleterious effects on Drosophila development associated with the sigma virus infection. Virology **22**, 142—148 (1964).
— Resistance to sigma virus infection in Drosophila. Nature (Lond.) **207**, 881—888 (1965).
— Low titers in Drosophila that regularly transmit the sigma virus infection. Genetics **53**, 79—83 (1966).
— Sigma virus content and hereditary transmission in *Drosophila melanogaster*. Virology **29**, 1—7 (1966).
SIGOT, A.: Contribution à l'étude de la sensibilité héréditaire au gaz carbonique chez la Drosophile. Influence de divers facteurs sur la transmission de la sensibilité par les mâles. Bull. biol. France Belg. **87**, 335—413 (1953).
VIGIER, PH.: Le problème de l'accumulation du virus infectieux dans une Drosophile sensible au gaz carbonique, stabilisée. C. R. Acad. Sci. (Paris) **252**, 2953—2955 (1961).
— Contribution à l'étude de l'instabilité génétique du virus σ de la Drosophile. Ann. Génét. **9**, 5—63 (1966).

Plant Pathogenic Viruses in Insects

KARL MARAMOROSCH*

With 6 Figures

Introduction

Until recently plant virus diseases have been studied mainly by plant pathologists, and the causative agents have been considered solely as disease agents of plants. Although the host range of most plant-pathogenic viruses is limited to plants, a small number of viruses are now known to infect not only plants but also insects that act as vectors and alternate hosts and reservoirs of such viruses. With the exception of a few viruses, transmitted through root grafts or pollen, all viruses that cause diseases of plants require intermediary carriers, so-called vectors, to pass from one plant to another. Such vectors are now known to occur among various orders of arthropods, among nematodes, and even among lower fungi (MARAMOROSCH, 1967).

Virus Multiplication in Vectors

In the past two decades it has been proven that several plant-pathogenic viruses multiply in their insect vectors and thus are actually not only plant, but also insect viruses. The reader is referred to detailed reviews of this subject (BLACK, 1953a, b, 1959; DAY, 1955; FUKUSHI, in press; MARAMOROSCH, 1953, 1954, 1955, 1958a, b, 1960, 1963, 1964, 1966, 1967; MARAMOROSCH and JENSEN, 1963; SMITH, 1957, 1958, 1962, 1965). The present chapter is primarily concerned with findings made within the past five years.

Most of our knowledge of the biological interaction between plant viruses and their insect vectors has come from the study of Cicadellidae, and recently also of Aphididae. The role of leafhoppers as vectors was discovered in Japan more than 70 years ago, and Japanese scientists pioneered in this type of work by demonstrating that some of the "plant viruses" can infect the progeny of leafhoppers congenitally, passing through the eggs to the offspring of vectors.

* Boyce Thompson Institute for Plant Research, Yonkers, New York.

Work by the author and his associates, Drs. R. R. GRANADOS, H. HIRUMI, E. SHIKATA and G. STREISSLE at the Boyce Thompson Institute was sponsored, in part, by U.S. Public Health Service Research Grants AI-04290 from the National Institute of Allergy and Infectious Diseases and CA-7453 from the National Cancer Institute, and by National Science Foundation Grants GB-1199 and GB-5444.

FUKUSHI (1933) discovered the transovarial infection of a rice leafhopper by the rice dwarf virus and carried out a transovarial serial passage that provided direct evidence for the multiplication of the virus in its insect carrier (FUKUSHI, 1935, 1940). BLACK (1944) discovered a clover-infecting virus that also had the property of infecting its leafhopper vector congenitally. This clover virus was used by BLACK (1950) to prove its multiplication in leafhopper vectors, by passing the virus through 21 generations of insects. The same method of transovarial passages has since been used by several others to prove multiplication of plant-pathogenic viruses in insect vectors (FUKUSHI, in press).

A different method of studying virus multiplication in insect vectors is by serial injection of virus, using mechanical inoculation, developed by STOREY (1933). The technique has since become a convenient method for bioassay of numerous plant viruses. The serial passage technique by needle inoculation was first applied to leafhoppers and plant viruses by MARAMOROSCH (1952), who used it to obtain direct evidence for the multiplication of aster yellows virus in its vector, *Macrosteles fascifrons*.

The multiplication of a plant pathogenic virus in insects can also be demonstrated by other methods. An assay of soluble wound tumor virus antigen was used by WHITCOMB and BLACK (1961), who measured the rate of increase of viral antigen in the leafhopper vector, *Agallia constricta*.

A method, recently devised by STROHMAIER (1967), has been used for assaying of the wound tumor virus, using heavy water gradient centrifugation and electron microscopy (STREISSLE et al., in press). A particle counting method, described by GÁMEZ and BLACK (1967), can also be used to ascertain virus multiplication in vectors. Finally, the examination of thin sections of vector cells sometimes permits the detection of virus assembly sites, so-called viroplasms, formed in cells in which the plant pathogenic virus multiplies (SHIKATA and MARAMOROSCH, 1967b). These viroplasms are similar to the so-called virogenetic network that has been identified as the site of granulosis insect virus multiplication (HUGER and KRIEG, 1961).

The accumulation of evidence for the multiplication of several plant-pathogenic viruses in leafhoppers has resulted in a search for similar relationships in other groups of insect vectors, and a search for possible deleterious effects of such viruses on their invertebrate animal hosts.

Deleterious Effects

Until 1956 it was generally thought that insect vectors of plant viruses were completely tolerant of the otherwise highly virulent plant viruses and that the long association between the insect host and the obligate virus parasite resulted in an almost symbiont-like relationship. The first evidence against this hypothesis was the finding of cytopathic changes in the fat body cell cytoplasm of aster leafhoppers, particularly males of *M. fascifrons* that were confined to virus-infected plants for prolonged periods (LITTAU and MARAMOROSCH, 1956, 1960).

The best documented case of a direct injurious effect of a plant-pathogenic virus on a leafhopper, and, in fact, the first such case of injury of any virus to its vector was made by JENSEN (1958, 1959) who demonstrated that the Western X-disease virus of stone fruits significantly shortened the life span of its vector, *Colladonus montanus*. More recently a very detailed cytological study was made by WHITCOMB et al. (in press) of the cytopathogenic and systemic effects caused by Western X-disease virus on *C. montanus*. Lesions were first detected in the epithelium of the first and second ventriculus. Four weeks after exposure changes were observed in the filter chamber and in the Malpighian tubules, while lesions in the first and second ventriculus became much larger and more numerous. While changes in the ventriculus sometimes became less acute with time, those in the Malpighian tubules were apparently irreversible and more severe.

The virus was found to be pathogenic to the leafhopper and the poor reproductive ability of infected females tended to eliminate the virus from colonies that were unable to renew their infectivity by feeding on infective plants. Several other leafhopper-borne viruses have since been found to be lethal to embryos, nymphs, or adult vectors. The resemblance of the relationship between the congenitally acquired Sigma virus of Drosophila (see Chapter No. 4 by SEECOF in this volume) and congenitally acquired leafhopper-borne viruses is apparent. However, Sigma virus is only pathogenic to fruit flies that are subjected to high concentrations of carbon dioxide. Similar treatment of leafhoppers infected with plant-pathogenic viruses does not seem to enhance virus pathogenicity.

Several leafhopper-borne plant viruses have been found to affect insect diapause and fecundity, and in some instances changes were found in the metabolism of virus-transmitting individuals (MARAMOROSCH and JENSEN, 1963).

Systemic Invasion

To illustrate the findings concerning the invasion of various organs of an insect vector by a plant pathogenic virus, the electron microscopy work carried out in the author's laboratory during the past 3 years with the leafhopper-borne wound tumor virus is briefly summarized below.

Ultrathin sections of viruliferous *Agallia constricta* VAN DUZEE, an efficient leafhopper vector and host of wound tumor virus, were prepared in an attempt to ascertain the sites of the virus within tissues and cells of various arthropod organs (SHIKATA and MARAMOROSCH, 1965a). Excised fat body tissues, gut, Malpighian tubules, and salivary glands, as well as whole abdomens, were fixed and embedded in epoxy resin. Alternating thick and ultrathin sections were made and stained with uranyl acetate and lead. For light microscopy, the thick sections were stained by DELAFIELD's hematoxylin and eosin. Thin sections were examined by a Siemens Elmiskop I electron microscope. The largest accumulations of wound tumor virus, often in the form of microcrystals, were found in the fat body tissues; crystallin arrangements were also seen in muscle cells, salivary glands, gut, nerve cell and hemocyte cytoplasm (Fig. 1). The nature of the microcrystal formation is not clear. It seems feasible that these micro-

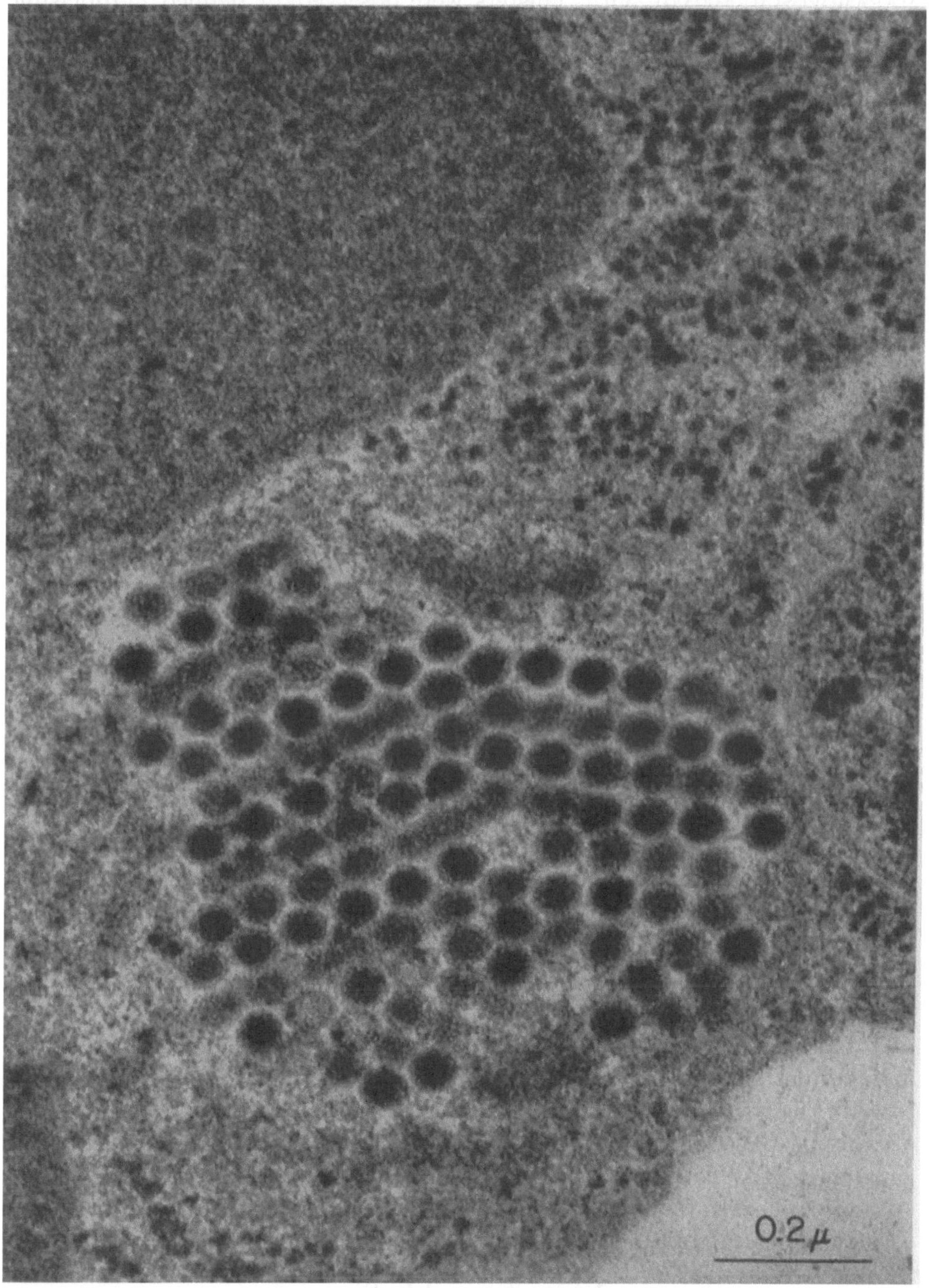

Fig. 1. Microcrystal of wound tumor virus in the cytoplasm of a hemocyte of the leafhopper vector *Agallia constricta*. × 110,000. (Electron micrograph courtesy of R. R. GRANADOS)

crystals represent the final stage of storage and accumulation of virus particles that had multiplied earlier in viroplasmic matrixes at other sites (SHIKATA and MARAMOROSCH, 1965b). Virus clusters were observed in defined cytoplasmic structures within the Malpighian tubules, mycetome, trachea and epidermis. Only one or two lobes of the salivary gland contained scattered virus particles not associated with defined cytoplasmic structures.

The findings pointed to the fat body as one of the likely sites of virus multiplication. The occurrence of virus accumulations in almost all organs of *A. constricta* provided direct evidence for systemic invasion of this arthropod host. Lack of a sizable accumulation of WTV in the salivary gland may explain the erratic and poor transmission record of the insect vector. This finding was in contrast with findings made with the very efficient leafhopper vector of the rice dwarf virus, *Nephotettix cincticeps*, in which the lobes of the salivary gland contain large masses of virus particles (FUKUSHI et al., 1962).

Various organs were also excised from the leafhopper *Agalliopsis novella* infected with wound tumor virus and were studied by electron microscopy in an attempt to determine the ability of an inefficient vector to support virus multiplication (GRANADOS et al., 1967). Small accumulations of virions were detected in the midgut epithelium, muscles, blood cells, fat bodies, tracheoblasts, and the ventral ganglia. Many small virus microcrystals were found in the ventral ganglia. Thus, the virus has been demonstrated to cause an infection in its arthropod host. Lack of sizable accumulations of virus not only in the salivary gland but also in other organs of *A. novella* might account for the inefficiency of this species as a vector of wound tumor virus.

Virus in Nervous System

The central nervous system of *Agallia constricta* infected with wound tumor virus was also studied by electron microscopy to obtain information concerning the virus distribution in the nervous system (HIRUMI et al., 1967). Wound tumor virions were found principally in the cytoplasm of the ganglion cells and less frequently in the glial cells. WTV was occasionally observed in the perineurium cells, nerve axons, tracheoblasts, and lateral nerves. In the ganglion cells, virions appeared as individual isolated particles in tubular formation and occasionally as aggregates (Fig. 2). In the glial cells, the virions were seen mostly in tubules, and very seldom in isolated or aggregated formations. In the perineurium cells and tracheoblasts, only small tubular formations were observed. The isolated virions were usually surrounded with polyribosomes and often appeared around the foci of the viroplasm. Sometimes degenerating ganglion cells infected with the virus were encountered. These damaged cells strongly indicated that the virus exerted a cytopathogenic effect on the nerve cells of the invertebrate host.

Virus Assembly Sites

Electron microscopy of thin sections of various organs of viruliferous insects indicates that the virus is assembled in accumulations of electron-dense aggregates, designated viroplasms, formed in the cytoplasm of fat body, muscle, midgut, Malpighian tubules, trachea, salivary gland, and epidermal cells (SHIKATA and MARAMOROSCH, 1967a). The viroplasms are considered the sites of

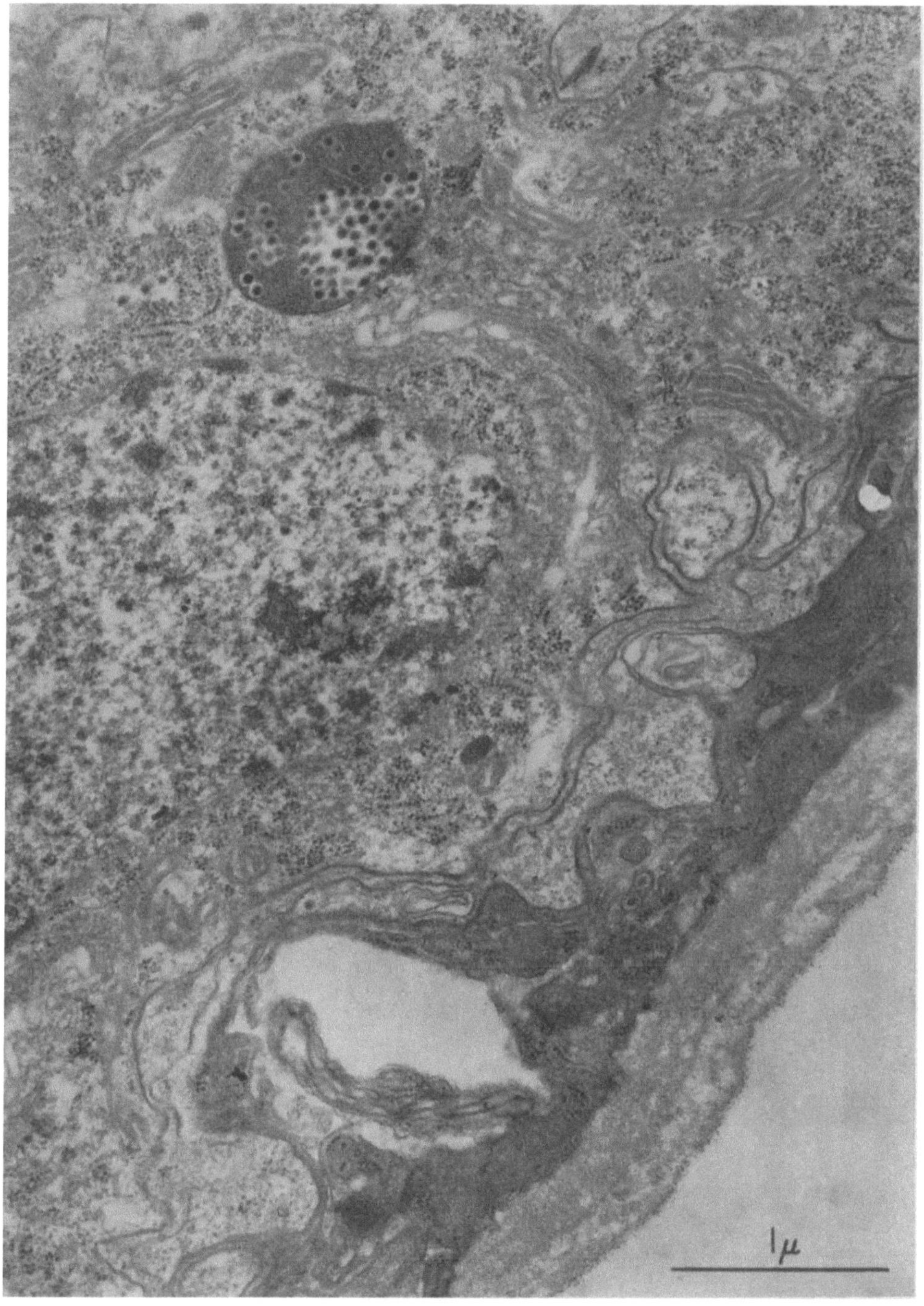

Fig. 2. Accumulation of wound tumor virus particles in the cytoplasm of a nerve cell of *A. constricta*. × 30,000. (Electron micrograph courtesy of R. R. GRANADOS)

virus assembly. Careful examination of infected plants during early stages of disease revealed the formation of viroplasms in the plant cells. Mature particles were found first at the periphery of the viroplasmic zone. A few days later vi-

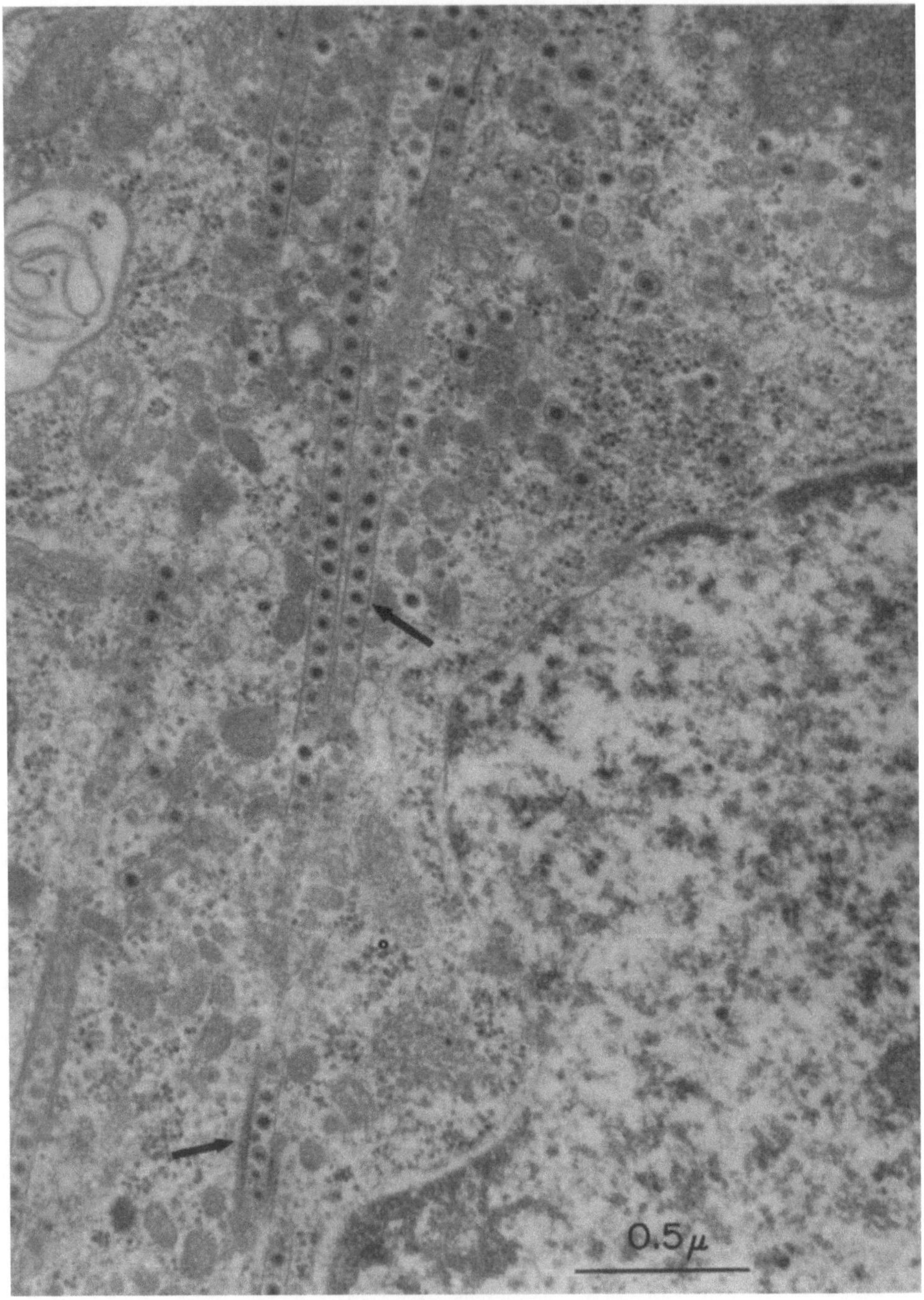

Fig. 3. Wound tumor virus within tubular structures (arrows), in the cytoplasm of a nerve cell of *A. constricta*. × 48,000. (Electron micrograph courtesy of R. R. GRANADOS)

rions were detected in increasing number inside the viroplasmic area. These morphological findings indicate that the wound tumor virus is assembled in plant and insect host cells in viroplasms formed within the cytoplasmic matrix

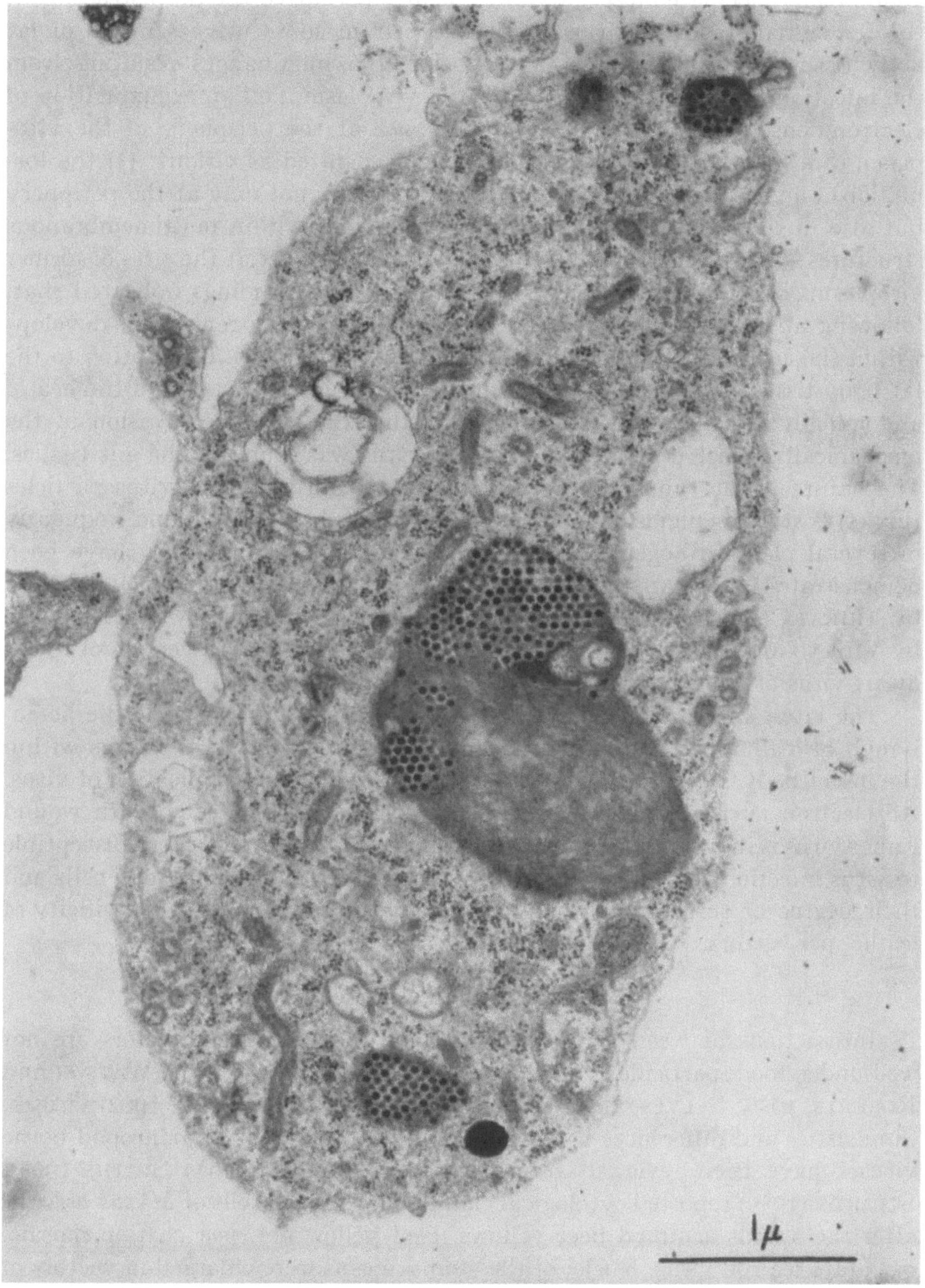

Fig. 4. Wound tumor virus in the cytoplasm of a hemocyte of *A. constricta*. × 27,000. (Electron micrograph courtesy of R. R. Granados)

proper. The similarity of wound tumor viroplasms to those of several viruses infecting higher animals and also to typical insect pathogenic viruses is striking.

The electron microscopy studies of the formation of wound tumor virus in abdominally inoculated insect vectors recently permitted the reconstruction of the sequential stages in virus development from consecutive samples of fat body tissues (SHIKATA and MARAMOROSCH, 1967b). Changes resulting from the infection comprised (1) formation of a viroplasm, i.e., an accumulation of electron-dense aggregates; (2) the appearance at the periphery of the viroplasm of a few fully formed virus particles recognized as virions; (3) the formation of an increasing number of individual virions, not only at the periphery but also in the matrix; (4) the engulfing of virions within multimembranous structures; (5) the formation of virus microcrystals either at the sites of former viroplasms or at some distance. These morphological findings indicated that, following abdominal inoculation of WTV, the plant-pathogenic virus develops within the cytoplasmic matrix proper of insect vector cells. In addition to the fat body tissues, WTV was detected in the epidermis, muscles, and trachea of abdominally inoculated insects, demonstrating the systemic invasion of the mechanically infected arthropod host. No virus was found in the gut tissues. The nature of the tubular structures that surround strings of virus particles (Fig. 3) is still an enigma. Such tubular structures have been found frequently in several plant-pathogenic viruses infecting insect vectors. They have been demonstrated by FUKUSHI et al. (1962), FUKUSHI and SHIKATA (1963), and by SHIKATA (1966) in rice leafhoppers that transmitted rice dwarf virus, and by VIDANO and BASSI (1966) and VIDANO (1967) in cicadellid vectors of rough dwarf virus of maize.

The latest findings of GRANADOS et al. (in press) indicate that the hemolymph of leafhopper vectors is not only the main carrier of the virus within the insect body, but also an important medium for the multiplication of virus. An electron micrograph of a blood cell of *A. constricta* infected with wound tumor virus is shown in Fig. 4. Only certain types of blood cells seem susceptible to virus infection. The presence or absence of susceptible hemolymph cells and their degree of susceptibility might explain the high biological specificity of leafhopper vectors.

Aphid-Borne Viruses

Interactions between plant-pathogenic viruses and aphid vectors are not well understood, particularly in the group of so-called persistent (WATSON and ROBERTS, 1939; SYLVESTER, 1962), circulative (KENNEDY et al., 1962) viruses. Similarities and differences between this and other groups of arthropod-borne viruses have been reviewed recently (MARAMOROSCH, 1964; SMITH, 1965). SCHMIDT (1959) reported cytological changes in stomach cells of *Myzus persicae* after the aphids acquired beet yellows, beet yellow net, pea enation mosaic, or potato leafroll virus. Nuclei of the stomach cells were enlarged in vectors of leafroll virus and of pea enation mosaic virus. Recently SHIKATA et al. (1966) examined ultrathin sections of pea plants and pea aphids exposed to pea

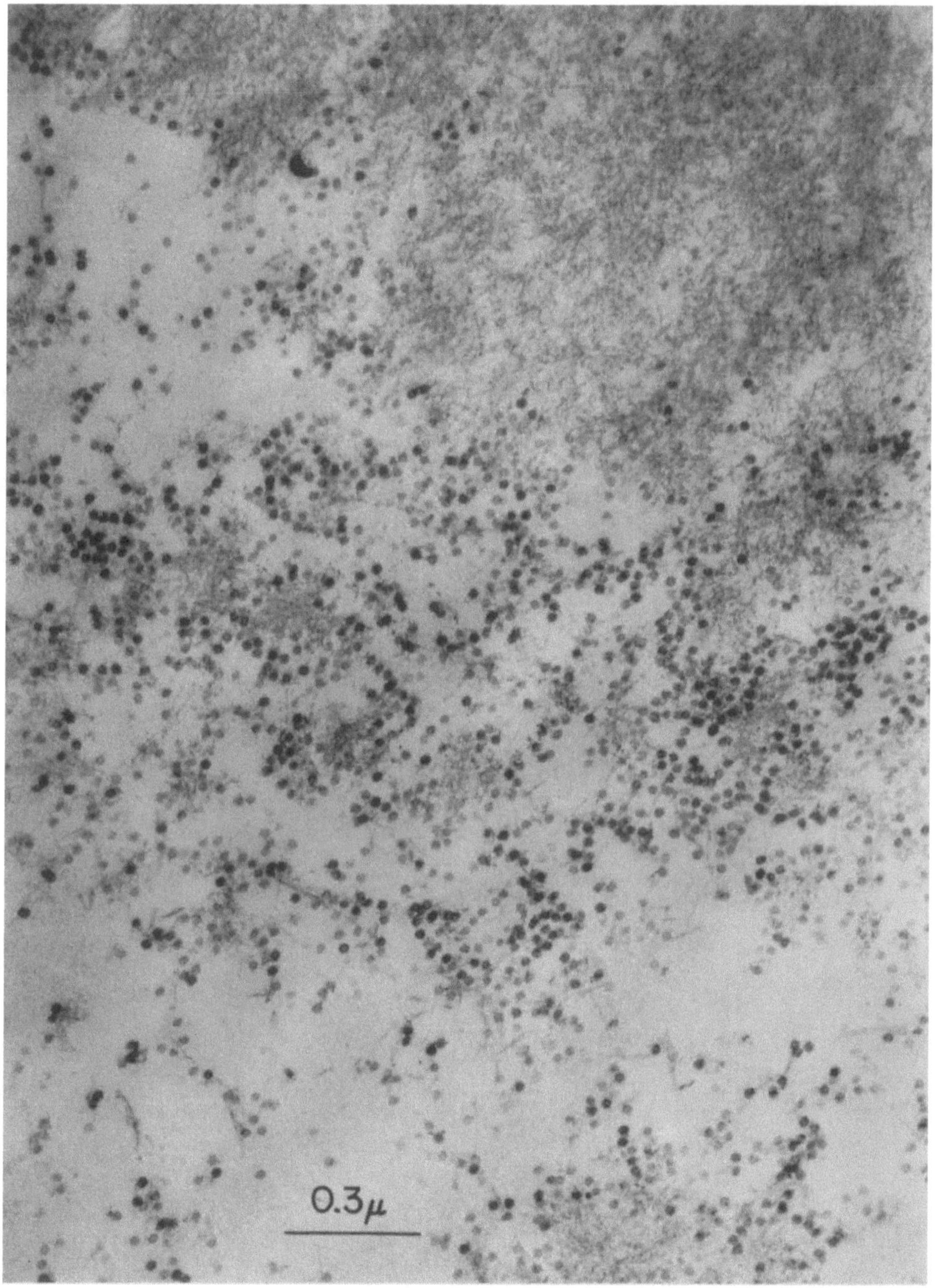

Fig. 5. Pea enation mosaic virus in the gut lumen of the aphid vector *Acyrthosiphon pisum*. × 60,000. (Electron micrograph courtesy of R. R. GRANADOS)

enation mosaic virus. Numerous spherical particles, about 28 millimicrons in diameter, were found in the cytoplasm and central vacuoles of necrotic and non-necrotic pea leaf and pod enations, but not in the mitochondria. Similar particles were detected in the gut lumen (Fig. 5) and fat body (Fig. 6) of

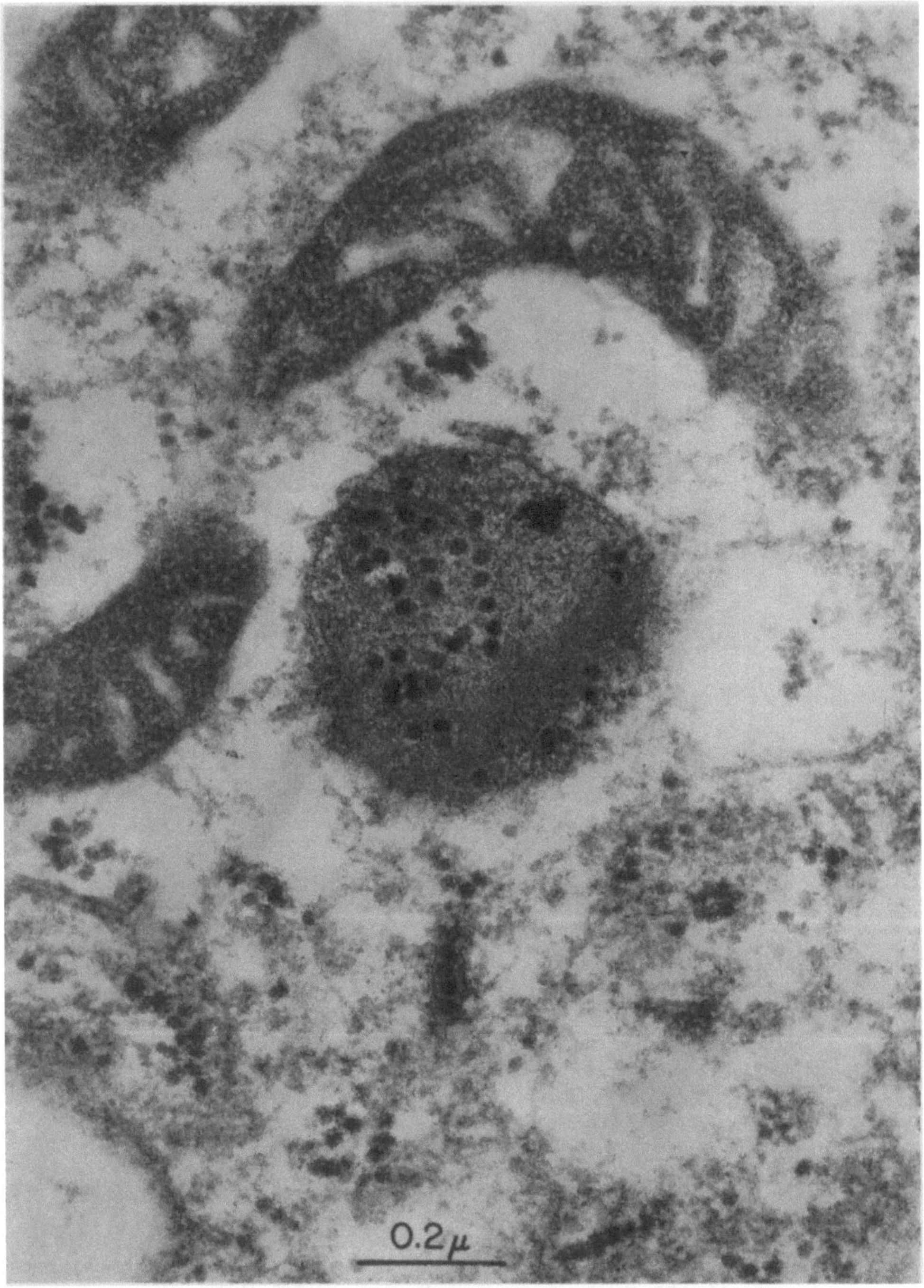

Fig. 6. Pea enation mosaic virus in the cytoplasm of a fat body cell of *A. pisum*. × 90,000. (From SHIKATA, et al., 1966.) (Reproduced from Virology 29, p. 430, by permission of Academic Press, Inc.)

viruliferous pea aphids. Dip preparations of sap from diseased plants also contained particles of the same size and shape. Similar particles were not found in healthy pea tissues, in virus-free pea aphids, or in dip preparations from healthy plant material. Pea enation mosaic virus is the first circulative aphid-borne virus localized by electron microscopy in plants and insects in situ.

The systemic infection of an aphid by a plant-pathogenic virus has been described recently by O'LOUGHLIN and CHAMBERS (1967). The authors found virions of lettuce necrotic yellows virus in *Hyperomyzus lactucae* aphids that fed on diseased plants and transmitted the virus. No particles were seen in aphids reared on virus-free plants or in embryos developing within infective vectors. Virions were detected in the muscles, fat body, brain, mycetome, tracheae, epidermis, salivary glands and in the cells of the alimentary canal. The authors concluded from their findings that the plant-pathogenic virus multiplied within the aphid.

The changes reported by SCHMIDT (1959) in nuclei of stomach cells in virus carrying aphids are not well understood, and it was of interest to establish whether pea enation mosaic was present in nuclei, or only in the cytoplasm of cells. To gain more information on the subject, it was decided to study the sequential invasion of various cell structures by this virus. In necrotic and non-necrotic cells of diseased plants, particularly in the earliest stages of the disease, large accumulations of virus particles were found in the nuclei. It appeared that the virus invades and then multiplies in plant cell nuclei, from which it penetrates into the cytoplasm of the infected cell. The study of aphid cells is still in progress and no data are available at this time on the possible presence of this virus in aphid cell nuclei.

Conclusions

Since it has now been established that several plant-pathogenic viruses do multiply in leafhoppers, and probably in other invertebrate vectors, and that in several instances the invertebrate hosts are adversely affected by the viruses, these circulative and propagative viruses cannot be considered merely as "plant" viruses. They are, in fact, both insect-pathogenic and plant-pathogenic. It may never be possible to establish whether such viruses originated in insects or in plants, but it is now certain that several leafhopper and aphid vectors of plant pathogenic viruses are not mere fortuitous carriers of viruses but hosts and reservoirs that may have played a primary role in plant and animal virus evolution. The effects of circulatory plant pathogenic viruses on insects suggest a symbiotic spectrum, from mutual benefit at one end to lethal parasitism on the other.

References

BLACK, L. M.: Some viruses transmitted by agallian leafhoppers. Proc. Amer. phil. Soc. **88**, 132—144 (1944).

— A plant virus that multiplies in its insect vector. Nature (Lond.) **166**, 852—853 (1950).

— Viruses that reproduce in plants and insects. Ann. N.Y. Acad. Sci. **56**, 398—412 (1953a).

BLACK, L. M.: Transmission of plant viruses by cicadellids. Advanc. Virus Res. **1**, 69—89 (1953b).
— Biological cycles of plant viruses in insect vectors. In: The viruses (F. M. BURNET and W. M. STANLEY, eds.), vol. 2, p. 157—185. New York: Academic Press 1959.
DAY, M. F.: Mechanisms of transmission of viruses by arthropods. Exp. Parasit. **4**, 387—418 (1955).
FUKUSHI, T.: Transmission of the virus through the egg of an insect vector. Proc. imp. Acad. Japan **9**, 457—460 (1933).
— Multiplication of virus in its insect vector. Proc. imp. Acad. Japan **11**, 301—303 (1935).
— Further studies on the dwarf disease of rice plant. J. Fac. Agr., Hokkaido Univ. **45** (pt. 3), 83—154 (1940).
— Relationships between propagative rice viruses and their vectors. In: Viruses, vectors and vegetation (K. MARAMOROSCH, ed.). New York: Interscience Publ. (in press).
—, and E. SHIKATA: Localization of rice dwarf virus in its insect vector. Virology **21**, 503—505 (1963).
— —, and I. KIMURA: Some morphological characters of rice dwarf virus. Virology **18**, 192—205 (1962).
GÁMEZ, R., and L. M. BLACK: Application of particle-counting to a leafhopper-borne virus. Nature (Lond.) **215**, 173—174 (1967).
GRANADOS, R. R., H. HIRUMI, and K. MARAMOROSCH: Electron microscopic evidence for wound-tumor virus accumulation in various organs of an inefficient leafhopper vector, *Agalliopsis novella*. J. Invertebrate Path. **9**, 147—159 (1967).
—, L. WARD, and K. MARAMOROSCH: Electron microscopy of leafhopper blood cells infected with wound-tumor virus. Bull. entomol. Soc. Amer. **13**, 199 (1967).
HIRUMI, H., R. R. GRANADOS, and K. MARAMOROSCH: Electron microscopy of a plant-pathogenic virus in the nervous system of its insect vector. J. Virology **1**, 430—444 (1967).
HUGER, A., and A. KRIEG: Electron microscope investigations on the virogenesis of the granulosis of *Choristoneura murinana* (HÜBNER). J. Insect Path. **3**, 183—196 (1961).
JENSEN, D. D.: Reduction in longevity of leafhoppers carrying peach yellow leaf roll virus (abstract). Phytopathology **48**, 394 (1958).
— A plant virus lethal to its insect vector. Virology **8**, 164—175 (1959).
KENNEDY, J. S., M. F. DAY, and V. F. EASTOP: A conspectus of aphids as vectors of plant viruses. London: Commonwealth Institute of Entomology 1962.
LITTAU, V. C., and K. MARAMOROSCH: Cytological effects of aster-yellows virus on its insect vector. Virology **2**, 128—130 (1956).
— — A study of the cytological effects of aster yellows virus on its insect vector. Virology **10**, 483—500 (1960).
MARAMOROSCH, K.: Direct evidence for the multiplication of aster-yellows virus in its insect vector. Phytopathology **42**, 59—64 (1952).
— Multiplication of a plant virus in an insect vector. Int. Congr. Microbiol., 6, Italy, 1953. Atti. **3**, 379—382 (1953).
— Biological transmission of plant viruses by animal vectors. Trans. N.Y. Acad. Sci. **16**, 189—195 (1954).
— Multiplication of plant viruses in insect vectors. Advanc. Virus Res. **3**, 221—249 (1955).
— Viruses that infect and multiply in both plants and insects. Trans. N.Y. Acad. Sci. **20**, 383—393 (1958a).
— Latent virus infections in arthropods. Int. Congr. Microbiol., 7, Stockholm, 1958. Symp. Recent Progr. Microbiol., p. 224—229 (1958b).
— Leafhopper-transmitted plant viruses. Protoplasma (Wien) **52**, 457—466 (1960).

MARAMOROSCH, K.: Arthropod transmission of plant viruses. Ann. Rev. Entomol. **8**, 369—414 (1963).
— Virus-vector relationships: vectors of circulative and propagative viruses. In: Plant virology (M. K. CORBETT and H. D. SISLER, eds.), p. 175—193. Gainesville: University of Florida Press 1964.
— Plant viruses. Int. Congr. Microbiol., 9, Moscow, Symp., p. 423—431 (1966).
— Transmission of plant pathogenic viruses. In: Transmission of viruses by the water route (G. BERG, ed.), p. 323—336. New York: Interscience Publ. (1967).
—, and D. D. JENSEN: Harmful and beneficial effects of plant viruses in insects. Ann. Rev. Microbiol. **17**, 495—530 (1963).
O'LOUGHLIN, G. T., and T. C. CHAMBERS: The systemic infection of an aphid by a plant virus. Virology **33**, 262—271 (1967).
SCHMIDT, H. B.: Beiträge zur Kenntnis der Übertragung pflanzlicher Viren durch Aphiden. Biol. Zbl. **78**, 889—936 (1959).
SHIKATA, E.: Electron microscopic studies on plant viruses. J. Fac. Agr. Hokkaido Univ. **55**, 1—110 (+ 255 Figs.) (1966).
—, and K. MARAMOROSCH: Electron microscopic evidence for the systemic invasion of an insect host by a plant pathogenic virus. Virology **27**, 461—475 (1965a).
— — Plant tumor virus in arthropod host: microcrystal formation. Nature (Lond.) **208**, 507—508 (1965b).
— — Electron microscopy of wound tumor virus assembly sites in insect vectors and plants. Virology **32**, 363—377 (1967a).
— — Electron microscopy of the in vivo formation of wound tumor virus in abdominally inoculated insect vectors. J. Virology **1**, 1052—1073 (1967b).
— —, and R. R. GRANADOS: Electron microscopy of pea enation mosaic virus in plants and aphid vectors. Virology **29**, 426—436 (1966).
SMITH, K. M.: Arthropods as vectors and reservoirs of phytopathogenic viruses. In: Handbuch Virusforschung **4**, 143—176 (1957).
— Transmission of plant viruses by arthropods. Ann. Rev. Entomol. **3**, 469—482 (1958).
— The arthropod viruses. Advanc. Virus Res. **9**, 195—240 (1962).
— Plant virus-vector relationships. Advanc. Virus Res. **11**, 61—96 (1965).
STOREY, H. H.: Investigations of the mechanism of the transmission of plant viruses by insect vectors. — I. Proc. roy Soc. B **113**, 463—485 (1933).
STREISSLE, G., V. BYSTRICKY, R. R. GRANADOS, and K. STROHMAIER: Number of virus particles in insects and plants infected with WTV. J. Virology (in press).
STROHMAIER, K.: A new procedure for the quantitative measurements of virus concentrations in crude preparations. J. Virology **1**, 1074—1081 (1967).
SYLVESTER, E. S.: Mechanisms of plant virus transmission by aphids. In: Biological transmission of disease agents (K. MARAMOROSCH, ed.), p. 11—31. New York: Academic Press 1962.
VIDANO, C.: Microecologia e moltiplicazione di MRDV nel vettore. Atti. Accad. Sci. Torino **101**, 717—733 (1967).
—, C., and M. BASSI: Dimostrazione al microscopio elettronico di particelle del virus del nanismo ruvido del mais (MRDV) nel vettore *Laodelphax striatellus* FALLÉN. Atti. Accad. Sci. Torino **100**, 73—78 (1966).
WATSON, M. A., and F. M. ROBERTS: A comparative study of the transmission of *Hyoscyamus* virus 3, potato virus Y and cucumber virus 1 by the vectors *Myzus persicae* (SULZ), *M. circumflexus* (BUCKTON), and *Macrosiphum gei* (KOCH). Proc. roy Soc. B **127**, 543—576 (1939).
WHITCOMB, R. F., and L. M. BLACK: Synthesis and assay of wound-tumor soluble antigen in an insect vector. Virology **15**, 136—145 (1961).
— D. D. JENSEN, and J. RICHARDSON: The infection of leafhoppers by Western X-disease virus. IV. Pathology in the alimentary tract. Virology **34**, 69—78 (1968).

A Review of the Use of Insect Tissue Culture for the Study of Insect-Associated Viruses

JAMES L. VAUGHN*

Insect Tissue Culture Systems

Viruses are associated with insects in a variety of host-pathogen relationships. The viruses may be specific pathogens of insects, as for example, are the polyhedrosis viruses; or they may be pathogens of higher animals or of plants, with the insects only the intermediate hosts and the vectors. Because of these diverse relationships, there is considerable interest among research workers in many areas of biology in the use of insect tissue culture as a tool for the study of viruses. Indeed the interest has often outdistanced progress in the field of insect tissue culture. Therefore, some discussion of several of the problems involved in the culture of insect tissue seems justified before reviewing the history and present status of the study of viruses.

Perhaps one of the most serious handicaps to the culture of insect tissue is the small amount of tissue available from each insect; for example, the ovaries from three or four *Bombyx mori* pupae are required to prepare a single 4 ml culture compared to 8 to 10 cultures of 500—700 ml from a pair of monkey kidneys. A method of assaying virus was described by VAUGHN (1963) in which he used only a quarter of an ovary per culture, but even this method provides only eight cultures from each insect. Several different tissues and methods have been tried by various workers in an effeort to overcome this problem. Undifferentiated embryonic tissues, for example, chick embryonic cells, are commonly used for vertebrate tissue cultures but the use of insect embryonic tissue presents several difficulties. HIRUMI and MARAMOROSCH (1964) and MITSUHASHI and MARAMOROSCH (1964) successfully established primary cultures of leafhopper embryonic tissue, but the embryos had to be removed from the eggs by tedious dissection. HORIKAWA and FOX (1964), however, cultured embryonic tissue of *Drosophila melanogaster* MEIGEN from preparations obtained by gently grinding surface-sterilized eggs in a glass homogenizer and primary cultures of embryonic cells of *Aedes aegypti* (L.) were prepared in much the same manner by PELEG (1966). The method may be suitable for preparing primary cultures

* Entomology Research Division, Agricultural Research Service, U.S. Department of Agriculture, Beltsville, Maryland.

of embryonic tissues whenever masses of eggs can be easily collected. However, our experience has shown that it is not suitable for eggs from most lepidopterous larvae because of the large amounts of yolk present.

The enzymatic treatment of masses of insect tissue to produce suspensions of cells has not been successful because trypsin, as normally used to dissociate mammalian tissue, damages insect cells (JONES, 1962). AIZAWA and VAGO (1959a) dissociated ovarian and epidermal tissue by partial digestion with trypsin followed by further dispersion with mechanical agitation. The yield consisted mostly of small clumps of cells rather than of individual cells, but further treatment with hyaluronidase produced a suspension of single cells that were used to establish primary cultures. MARTIGNONI et al. (1958) found that hydrolytic enzymes obtained from the hepatopancreas and crop of the snail *Helix aspersa* MÜLLER would disassociate the thoracic tissues of the variegated cutworm *Peridroma saucia* (HÜBER) to give satisfactory suspensions of undamaged, single cells. However, this method has not been tried by other workers.

Insect blood cells, like the blood leukocytes in mammals retain the ability to divide *in vitro*. They are susceptible to many insect viruses, are abundant, and are easy to collect. However, the blood cells of many insects contain polyphenol-oxidase which, in the presence of oxygen produces several toxic quinone substances. MARTIGNONI and SCALLION (1961a) blocked the action of this enzyme by adding ascorbic acid to their medium and by culturing the cells of *P. saucia* in an atmosphere of 96% nitrogen, 1% oxygen, and 3% carbon dioxide. Although the cell attached and formed a monolayer, mitosis was not demonstrated. The first successful long-term cultivation of blood cells was accomplished by MITSUHASHI (1966a) using cells from the rice stem borer, *Chilo suppressalis* (WALKER). These cells had been in continuous culture for 12 months and had gone through eight serial subcultures. Since MITSUHASHI describes no special precautions to prevent intracellular melanization, one must assume this was not a problem. Thus, in those insects where intracellular melanization does not occur, blood cells are an excellent source of material for *in vitro* virus studies.

Continuously propagatable cells such as are available with vertebrate cell lines would be of great value to virologists working with insect viruses because of the difficulties, just discussed, in obtaining primary cultures. Two such insect cell lines are now in existence — one from a moth *Antheraea eucalypti* SCOTT [*Austrocaligula eucalypti*] (SCOTT) (GRACE, 1962a) and the other from the mosquito *Aedes aegypti* (GRACE, 1966). These two cultures have been widely distributed, but their use has been restricted because hemolymph is needed as a supplement in the medium. Recently, a subline of the *Antheraea* line was developed (YUNKER et al., 1967) that grows on a medium supplemented only with bovine sera and egg ultrafiltrate and is not dependent on insect hemolymph. This subline should be useful to investigators who do not have access to supplies of insect hemolymph.

The second major problem encountered by investigators who wish to study viruses in insect tissue culture is the choice of a medium. Since extensive reviews of this subject were published by JONES (1962) and by MARTIGNONI (1962) only the more important aspects will be discussed here.

Over the years, investigators have found that generally, insect hemolymph differs in composition from mammalian blood in: the broad range of the sodium-potassium ratio (0.1 to 25), the higher level of calcium and magnesium ions, the increased level of amino acids, the presence of organic phosphates, the presence of organic acids, the presence of the disaccharide trehalose and a pH between 6.4 and 6.7. WYATT (1956) developed the first medium based on this accumulated information. Designed for tissue from phytophagus insects, this medium has a mole sodium-potassium ratio of 0.2. GRACE (1962a) modified this medium by adding most of the B-vitamin complex and raising the sucrose level from 40 mg/ml to 2668 mg/ml. The modified medium is now commercially available and I have found it suitable for the primary culture of tissue from many phytophagous insects. When supplemented with hemolymph from *Antheraea pernyi* (GUÉRIN-MÉNEVILLE), it is used to culture the GRACE *Antheraea* and the GRACE *Aedes* lines. Other workers have substituted yeast extract and lactalbumin hydrolysate for the B-vitamin complex and the amino acids, for example, VAGO and CHASTANG, 1958; JONES and CUNNINGHAM, 1961; HIRUMI and MARAMOROSCH, 1964; MITSUHASHI, 1965a.

Since the hemolymph of zoophagous insects generally has a mole ratio of sodium to potassium greater than 1 and omnivorous insects a mole ration that varies but is around 1, media prepared for vertebrate tissue culture has been used at least in part for the culture of tissue from these insects. REHACEK (1963) used equal parts of EAGLES' and VAGO and CHASTANG's media supplemented with 5% dextran and 0.1% of an unidentified calf serum fraction to culture explants of adult tissue from *Dermacentor marginatus* (SULZAR) and *Dermacentor pictus* (HERMANN). TING and BROOKS (1965) reported no outgrowth from explants of tissue from the cockroach, *Blattella germanica* L. unless the mole ratio of sodium-potassium exceeded 1.

Most insect cultures required serum supplements in the medium for satisfactory results. Often this supplement is the hemolymph of the insect species from which the tissue is taken though it need not be. In our laboratory, hemolymph from the silkworm, *B. mori*, has been used in media for the culture of tissue from several insects including *Galleria mellonella* (L.), *Malacosoma americanum* (F.), and *Estigmene acrea* (DRURY). SEN GUPTA (1964) tested hemolymph from *G. mellonella*, *A. pernyi* and *Melolontha melolontha* (L.), respectively, as medium supplements for tissue from each of the same three insects and found no specificity in the hemolymph. Other studies have shown that the entire serum supplement need not be hemolymph. VAGO and CHASTANG (1962) reported that calf serum could be used to replace as much as two-thirds of the insect hemolymph in primary cultures of tissue from several Lepidoptera and fetal bovine serum has been used successfully as a partial replacement for hemolymph in both primary cultures and cell line cultures in our laboratory. PELEG (1966) used 2% chick embryo extract as the only supplement in his medium for the culture of embryonic *A. aegypti* cells. Thus, insect hemolymph is still a necessary supplement, but it need not come from the same insect as did the tissue, when that is impractical, for example, because of the size of the insect. In many cultures, some of the more easily obtained vertebrate sera may be substituted for hemolymph to some degree.

The preceding discussion is by no means a complete report on the culturing of insect tissue. The topics were chosen to illustrate aspects that are pertinent to the use of insect tissue cultures for the study of viruses. The references were chosen to illustrate attempts to culture tissue from insects that have some special interest to virologists or because the paper contains an extensive review of the literature. It is intended to provide a background for the review of the *in vitro* studies of insect-associated viruses.

Studies involving Viruses Pathogenic to Insects

Methods of Infection

Insect tissues in culture can be infected with virus in at least two ways; one, with hemolymph collected from diseased insects and two, with purified preparations of the viral particles. To obtain particle suspensions of the inclusion body viruses the inclusion bodies must be dissolved with dilute Na_2CO_3. Intact inclusion bodies are not infective as they can only be dissolved under conditions of reduced oxidation-reduction potential and at a pH of 9 to 11, conditions that are not compatible with tissue culture.

The first successful infection of insect tissue *in vitro* by an insect virus was reported by TRAGER (1935). He inoculated primary cultures of ovarian tissue from *B. mori* with dilutions of hemolymph from diseased larvae. Typical polyhedra appeared in the nuclei of the cells in both the outgrowth and the original explant. Inoculation of silkworm larvae with the culture medium indicated infectivity in the cultures had increased by more than 1000 times over the original level. The virus was also transmitted serially in tissue culture by transferring small pieces of infected tissue to healthy cultures and by transferring medium from infected cultures to new cultures. The virus was passed serially by medium for seven passages, and although the inoculum in the last cultures represented a 1.56×10^{-14} dilution of the original material, infections were still produced in the cultures. Thus, TRAGER early established that insect tissue cultures would produce infectious particles and typical polyhedra when they were inoculated with insect virus.

Some years later GRACE (1958) reported the unexpected appearance of polyhedra in tissue cultures from the tussock moth, *Hemerocampa leucostigma* (J. E. SMITH) used in tests of medium with and without hemolymph. In cultures started on medium without hemolymph, polyhedra developed when hemolymph from the promethia moth, *Callosamia promethea* (DRURY) was added to the medium. The polyhedra produced were infectious for the tussock moth (GRACE, 1959) but not for the promethia moth and GRACE concluded that the tissue had a latent infection that became manifest after the abrupt change in media.

AIZAWA and VAGO (1959b) published a confirmation of TRAGER's early work. They also were successful in infecting cells of *B. mori* ovarian tissue *in vitro* with the silkworm nuclear polyhedrosis virus by using hemolymph from diseased larvae and also by using medium from infected cultures. In addition, they showed that explants from silkworm in an early stage of the disease supported the completion of the virus development *in vitro*, suggesting that this might be a way of demonstrating the presence of virus in insects.

Similar results were published simultaneously by GAW et al. (1959), who also confirmed that these two methods were satisfactory for producing the silkworm nuclear polyhedrosis virus in silkworm ovarian culture. Whereas TRAGER, and later AIZAWA and VAGO had used new primary cultures, GAW and his group demonstrated that gonadal cells carried through 22 successive generations remained normal in appearance and showed typical cytopathic changes when infected with the nuclear virus.

The use of hemolymph from diseased insects was subsequently shown to be an effective method for infecting cultures of cells from other insects with the corresponding nuclear viruses. MARTIGNONI and SCALLION (1961b) reported the infection of cultures of blood cells from the varigated cutworm *P. saucia* with the nuclear polyhedrosis virus of this insect. Also HEITOR (1963) and SEN GUPTA (1963) reported the production of normal polyhedra in cultures of ovarial cells, fat body cells and blood cells from *G. mellonella* when they were inoculated in this manner.

Despite the success obtained when hemolymph was used to inoculate insect tissue cultures the use of suspensions of virus rods obtained by dissolving the polyhedra in dilute alkali has many advantages, one being the ease of preparation. The polyhedra can be washed clean by repeated centrifugation and stored dry for long periods, thus diseased insects need not be on hand continually. Instead, preparations of the virus rods, free of contaminating material can be quickly prepared from weighed amounts of polyhedra. A second advantage is that for quantitative studies particle counts can be made. Such counts are difficult to make from insect blood because particles of a similar size normally occur in healthy blood. However, VAGO and BERGOIN (1963) reported that the infection of tissue cultures with suspensions of virus rods was rarely possible. Simultaneously, VAUGHN and FAULKNER (1963) who were attempting to develop assay methods for the *B. mori* polyhedrosis virus in tissue culture found that virus rods separated from polyhedron protein were not infectious. When several series of primary ovarian tissue cultures were inoculated with suspensions of virus rods removed from the polyhedron protein no cultures became infected though the virus concentration in these suspensions was equivalent to 10^6 LD_{50} for insects. To determine if the virus rods were altered in some manner by the treatment necessary to remove them from the polyhedra, hemolymph from diseased insects was treated in the same way, i.e., the pH was adjusted to 10.5 with dilute Na_2CO_3, incubated at room temperature for 1 hour and then added to primary tissue cultures. These tests demonstrated that the treatment did not alter the infectivity of the hemolymph. To test whether or not hemolymph might alter the virus particles in some way to increase the infectivity the virus particles were preincubated in hemolymph from healthy insects. When tissue cultures were inoculated with the virus-hemolymph mixtures no infection resulted, but when the virus rods were mixed with dissolved polyhedron protein, (shown not to be infectious alone) and used to inoculate the cultures, infection did result. The authors concluded that the polyhedron protein contributed, in some unexplained way, to the infectivity of the virus rods.

Later FAULKNER and VAUGHN (1965) found that silkworm primary cultures could, in fact, be infected with the virus rod suspensions alone. However, the infectivity was considerably lower for tissue culture than for insects. Virus suspensions prepared from 17.5 mg of dry polyhedra dissolved in 0.05 M Na_2CO_3, washed by centrifugation and resuspended in 5 ml of phosphate buffered saline solution normally contained approximately 6×10^8 LD_{50} units/ml when titrated in insects. Titration of the same preparations in tissue culture indicated only 2×10^3 LD_{50} units/ml. In comparison, values obtained when a single sample of hemolymph from diseased insects was titrated in the two systems were 1×10^8 LD_{50} units/ml in insects and 4×10^6 LD_{50} units/ml in tissue culture. Thus, the two values for a virus rod preparation differed by several hundred thousand LD_{50} units/ml, corresponding values for hemolymph differed by only about a hundred LD_{50} units/ml.

No explanation for the difference is yet known. However, VAUGHN (1965) made attempts to purify the infectious material in the hemolymph using a CM-Sephadex[1] cation-exchange column. He was able to show that the hemolymph from diseased insects routinely had two peaks of activity: the material in the first was not adsorbed on the exchange column and was recovered by eluting the column with 0.01 M phosphate buffer at pH 7.8. The material in the second was recovered from the column when the ionic strength of the eluent was increased by the addition of 0.4 M KCl. Chromatography of either a virus rod suspension separated from the polyhedron protein or the virus rods plus the dissolved polyhedron protein gave only one peak, corresponding to the first peak obtained from hemolymph. Thus a demonstrable difference exists between viral preparations from hemolymph and those from polyhedra dissolved in dilute Na_2CO_3. Whether this difference reflects the difference in the virulence of the two preparations in tissue cultures will not be known until more is known about the infectious material in the hemolymph of diseased insects.

Insects are susceptible to a second group of polyhedral viruses, the cytoplasmic polyhedrosis viruses. The polyhedra of these viruses form in the cytoplasm of the cells instead of in the nucleus. Also they are spherical rather than rod-shaped and contain RNA instead of DNA as the nucleic acid component. The cytoplasmic polyhedrosis viruses have not been studied in any detail in tissue cultures.

---

[1] Mention of a company name or proprietary product does not necessarily imply endorsement of the company or product by the U.S. Department of Agriculture.

GRACE (1962b) reported the development of a cytoplasmic virus in cultures of tissue from *A. eucalypti* after they had been inoculated with the medium from primary cultures of *B. mori* infected with the nuclear polyhedrosis virus of *B. mori*. He also reported a spontaneous appearance of cytoplasmic virus in his cultures of *Antheraea* cells after they had been grown *in vitro* for about 10 months. When fluid portions of cultures of *B. mori* containing nuclear polyhedra were fed to larvae of *A. eucalypti* and *B. mori*, cytoplasmic virus developed in 1 of 5 *A. eucalypti* larvae and nuclear virus developed in the *B. mori* larvae. When the polyhedra were removed from the medium and the medium was fed to groups of larvae from each species, no disease developed. (This result was not unexpected since only polyhedra and not free virus are infectious to larvae *per os*). When medium from *Antheraea* cultures infected with cytoplasmic virus was used to inoculate tissue cultures from each species or fed to larvae of each species he reported the following results: cultures and larvae of *A. eucalypti* developed cytoplasmic virus; cultures of *B. mori* were not infected; larvae of *B. mori* developed both cytoplasmic and nuclear virus. GRACE concluded that the tissue of *A. eucalypti* contained a latent infection that was induced to develop in tissue culture.

The nuclear viruses are species specific, but the cytoplasmic viruses have been shown to be less so (SMITH, 1967). Therefore, if the original material from the *B. mori* cultures contained a small amount of cytoplasmic virus as well as the known nuclear virus, the events described by GRACE would have occurred as a consequence of this undetected mixed infection. Since GRACE did not definitely rule out the possibility of mixed infection it is perhaps as possible as latency. A year later, VAGO and BERGOIN (1963) established that material obtained from the intestinal cells of *Porthetria dispar* infected with cytoplasmic polyhedrosis virus would produce infection in cell cultures, but no attempts have been reported in which the isolated particles of this virus have been used as an inoculum.

The third type of inclusion body viruses affecting insects, the granulosis virus, also develops in the cytoplasm of the infected cell. However, granulosis virus differs from the other two because each inclusion body contains only a single virus particle. The only report of the multiplication of this virus in tissue culture was by VAGO and BERGOIN (1963). When fatty tissue of larvae infected with granulosis virus was used to inoculate ovarian cultures, inclusion bodies were formed that had the form and size of the granules normally associated with this virus and smears of the cells examined with the electron microscope confirmed the presence of granules and of the virus rods in the granules.

Many viruses infecting insects do not form the typical protein inclusion bodies. Of this group, only the iridescent viruses have been shown to multiply in tissue culture. In 1964, BELLETT and MERCER, in the first of a series of papers, reported that the *Sericesthis* iridescent virus (SIV) had been grown in the cells of the GRACE *Antheraea* line inoculated with a purified virus suspension. It was shown that one infective unit initiated a single cytoplasmic focus of antigen and thus the effective multiplicity of infection could be checked by staining with fluorescent antibody and counting the number of antigenic foci within infected cells. Typical virus development occurred when the cultures were inoculated with about two *in vitro* infectious units per cell. The proportion of infected cells could not be increased by using a larger inoculum, and, in fact, decreased when multiplicities of 10 or more infective units per cell were used. However, infection was affected by the temperature at which the cells were incubated: cultures incubated at $21° \pm 1°$ C required about one-tenth as much virus to infect the cells as cultures incubated at 25°.

Later BELLETT (1965 a) proposed an assay system for SIV in the *Antheraea* cell line: the virus titer in infective units/ml was estimated from the proportion of cells that were stained with fluorescent antibody 6—1/2 days after inoculation with dilutions of the virus. At multiplicities below 0.5 infective units/cell, the proportion of uninfected cells was nearly that predicted by the Poisson distribution. The particle-infectivity ratio was about 80:1 for purified virus, an indication that the *in vitro* system was about 10 times less sensitive than that reported for an *in vivo* system.

There have been two other reports recently of the infection of tissue cultures with purified virus preparations. In the first, MITSUHASHI (1966b) reported development of *Chilo* iridescent virus in primary cultures of hemocytes, fat body cells, gonadal tissue and brain of the rice stem borer, *C. suppressalis* after inoculation with purified suspensions of the virus obtained from diseased *C. suppressalis* larvae. In the second report VAGO et al. (1966) described the development of a nuclear noninclusion virus in primary cultures after inoculation with a purified virus preparation obtained from diseased larvae of *G. mellonella*.

The problems of investigators attempting to inoculate insect tissue cultures with purified virus preparations of the polyhedrosis viruses were not encountered by the investigators cited above, who used the noninclusion viruses. The *in vitro* assay developed by BELLETT (1965 a) for the iridescent virus was shown to be only slightly less sensitive than an *in vivo* assay.

Host Specificity *in vitro*

Researchers have known for some time (LURIA, 1963) that viruses affecting vertebrates do not always exhibit the usual pattern of host specificity when grown in tissue culture. For example, vaccinia, yellow fever, influenza and many other viruses grow in a chick embryo culture although they will not grow in chickens. Among the insect viruses, the nuclear polyhedrosis virus is generally considered the most host specific: cross-infection between hosts generally occurs only under experimental conditions when very large doses of polyhedra are given. The granulosis viruses also are highly host specific, no instance of successful cross transmission has been demonstrated by *per os* inoculation (HUGER, 1963). However, since cytoplasmic polyhedrosis virus and the irisdescent viruses are generally not species specific and infection of insect species other than the original is common, the development of these viruses in cell cultures from insects other than the host species would be expected. Thus,

BELLETT and MERCER (1964) and BELLETT (1965a and 1965b) studied the *Sericesthis* iridescent virus originally isolated from a beetle, *Sericesthis pruinosa* DALMAN, in the cultures of a cell line obtained from *A. eucalypti*.

Lack of host specificity for the nuclear polyhedrosis viruses was indicated by the work of VAGO and BERGOIN (1963) who reported that primary cultures of gypsy moth tissue (*P. dispar*) were readily infected with the nuclear polyhedrosis viruses from *B. mori* and *A. pernyi*. Although they made no back feeding tests to demonstrate pathogenicity for the original hosts, polyhedra which developed in the cultures retained the triangular shape usually associated with the virus of *A. pernyi* but rarely with the virus of *P. dispar*. Also VAGO et al. (1966) reported that a recently discovered noninclusion nuclear virus pathogenic for the wax moth developed normally in primary ovarian cultures from *B. mori* pupae.

Thus far, host specificity of insect viruses *in vitro* has been studied somewhat superficially, in only a few insects, and with only a few viruses; therefore, no general conclusions can be drawn. It is important to note that although the *Antheraea* cell line has been available for five years and has been distributed to many laboratories only work of BELLETT and MERCER, 1964; and BELLETT, 1965a and 1965b, have reported the growth of an insect virus in these cells.

Tissue Specificity *in vitro*

In insects, the nuclear polyhedrosis viruses multiply in the epidermis, tracheae, fat body, and blood cells. During the later stages of infection, polyhedra may form in the cells of the silk glands, nerve ganglia cells and imaginal buds (SMITH, 1967). In our laboratory we have several times observed polyhedra in sections of ovaries. Since these are the tissues most commonly used in insect tissue culture it is not surprising that many workers report that these viruses grow well in these tissues *in vitro*. For example, development of the nuclear polyhedrosis viruses was reported in cultures of blood cells by MARTIGNONI and SCALLION (1961) and in cultures of fat body cells by SEN GUPTA (1963). TRAGER (1935), VAGO and CHASTANG (1960), and many others have reported multiplication of these viruses in cultures of ovarian tissue from different insects. Recently in our laboratory we examined stained sections of cultured ovaries from late fifth instar *B. mori* and found that 10 days after the cultures were inoculated polyhedra had developed in all tissues of this organ, including the follical cells (unpublish data).

The iridescent viruses also develop in many tissues of an insect. Initially these viruses multiply in the fat body; as the disease progresses muscle, wing buds, legs and head may become infected (SMITH, 1967). MITSUHASHI (1966b) showed that primary cultures of gonadal tissue, circulatory organs, fat body cells, and brain were suitable for virus development.

The cytoplasmic polyhedrosis viruses develop first in the epithelial cells of the midgut. As the disease progresses, the polyhedra develop in cells of the foregut and hindgut (SMITH, 1967). They are not known to develop in any other cells in the insect. However, GRACE (1962) demonstrated the development of cytoplasmic polyhedra in primary cultures of ovarian tissue and material from these cultures was infectious for both the primary ovarian cultures and larvae of the same species. Also, VAGO and BERGOIN (1963) reported that midgut material from larvae infected with cytoplasmic virus was capable of infecting primary cultures of ovarian tissue. Thus the cytoplasmic polyhedrosis viruses, though they have a definite tissue specificity *in vivo*, are apparently capable of infecting other tissues *in vitro*.

Virus Development *in vitro*

The development of a virus in the host cell can be divided into three general phases, each of which involves several steps. The first phase is the entry of the virus into the host cell, the second is the development of new virus material in the cell and the third phase is the release of the newly formed virus. Extensive studies on sections of diseased insects with the electron microscope have provided a static picture of what happens during the second phase, but nothing has been discovered about the first or the third phases. All of these phases are, of course, amenable to study in tissue culture. However, only a few studies, other than with the light microscope, have been made of these phases of the host-virus interaction.

The first interaction between a virus and the host cell is the attachment to, or entrance into the cell by the virus. AIZAWA (1959) reported adsorption of the nuclear polyhedrosis virus to cells maintained in suspension in a medium that provided maintenance but not growth of cells. The amount of virus in the fluid portion of the tissue culture decreased for several hours after inoculation. To establish that the decrease was the result of adsorbtion to the tissue, he incubated dilutions of hemolymph from diseased insects with and without tissue at 15° and 12° C for 12 hours. The medium without tissue had an LD_{50} of $10^{-4.30}$ and $10^{-4.55}$ respectively; the medium with tissue had an LD_{50} of $10^{-2.93}$ and less than $10^{-2.30}$ respectively. When physiological saline solution was used instead of medium no differences were detected in the titers of the two systems at either temperature, thus no adsorbtion had occurred. To test the effect of non-insect tissue, AIZAWA substituted chick embryo cells for the insect cells, again no adsorbtion occurred since there were no differences in titers with or without cells at either temperature.

Thus adsorbtion of the NPV to insect cells does take place as was previously indicated by loss of infectivity in insect tissue cultures. Also, as with viruses of higher animals, adsorbtion of insect viruses was not inhibited by temperatures that were low enough to prevent or at least delay multiplication of the virus. Also some processes or sites in susceptible insect tissue but not in other tissues (at least not in chick embryo tissue) were involved. AIZAWA interpreted the results of his tests with physiological saline solution to mean that unknown substances are required if the adsorption is to occur. He did not describe the effect of the saline solution on the cells, but the hypotonic solution probably caused some damage and this damage could also explain all or some of the lack of adsorption.

The cellular changes that occur as the nuclear polyhedrosis viruses develop in tissue culture were described by TRAGER (1935). Subsequent descriptions of the *in vitro* development of several polyhedrosis viruses are very similar '(AIZAWA and VAGO, 1959; VAGO and CHASTANG, 1960; MARTIGNONI and SCALLION, 1961 b; VAGO and BERGOIN, 1963; and SEN GUPTA, 1963). Also ,then changes that occur *in vitro* are similar to those observed in hemocytes during development of the virus *in vivo*.

Electron microscopic observations of the development of a nuclear polyhedrosis virus in tissue culture cells was reported by VAGO and CROISSANT (1962). Primary cultures of *B. mori* ovarian tissue were cultivated on small plastic discs and inoculated with hemolymph from diseased larvae 36 hours after the explants had attached to the surface. At 12-hour intervals, a disc was removed, and the cells fixed in a veronal-osmic acid fixing agent at 4° C for one hour. Then the cells were dehydrated with a series of increasing concentrations of ethyl alcohol and detached from the disc. The free cells were concentrated by centrifugation, and the pellet was embedded in a mixture of butyl and methyl methacrylates. Sections were made with a PHILIPS and REICHART ultramicrotom.

The first pathological signs were observed after 24 or 36 hours when the mitochondria and the vacuoles near the nuclear membrane appeared somewhat distorted. Some hypertrophy of the nucleaus was perceptible, and a few electron dense regions appeared near the nucleoli. No viral elements were seen. After 48 hours, the infected nuclei were considerably enlarged, and by 60 hours after inoculation, they had developed clear areas in which small, slender rods, 25 to 30 mμ in diameter and 50 to 200 mμ in length were seen. At 72 hours, extensive clearing had occurred in the infected nuclei, and the rod-shaped bodies were more numerous, occurring not only in the cleared areas, but throughout the nuclear sap. Larger rods were also present which when cut transversely or obliquely, appeared as structures having a dense core, about the diameter of the earlier structures and surrounded by a less dense material, that did not appear to be a membrane. Presumably these structures were the completely formed virus rods.

After four days of culture, the nuclei contained a large number of the virus rods and early forms of polyhedra that lacked the typical sharp edges. Virus rods were seen in all stages of incorporation in the inclusion body. In the last stages, the bulk of the inclusion body protein was in the process of closing around the virus particles, all of which were by this time completely embedded in the inclusion body. Only then did the inclusion body appear to have the normal structure. At the end of five to six days, the nuclei appeared to contain only completely formed polyhedra, and free virus rods were not seen at this stage except where they were associated with what appeared to be amorphous debris.

In later stages of development, the nuclear membrane contained breaks through which the polyhedra were released into the cytoplasm. In these cells, the structure of the cytoplasm was greatly changed; mitochondria were no longer visible and vacuoles were extremely numerous. Just before the cell disintegrated the cytoplasm appeared as a vacuolar structure with little substance visible between the vacuoles.

Cells in culture normally contain fewer but larger polyhedra than do cells in insects, which suggests that the size of the polyhedra is determined by the metabolism of the cell (VAGO and CHASTANG, 1960). There is a greater variation in the shape of polyhedra in tissue culture than is normally seen in insects, for example, in *B. mori* cultures infected with the nuclear polyhedrosis virus, the typical polyhedral shape is hexagonal, but cuboidal and triangular polyhedra are often seen in tissue culture. Adjacent cells may contain different shaped polyhedra, but all polyhedra within a single cells are the same shape, which led us to speculate that the shape might be genetically controlled by the virus and that only a single particle is responsible for the initiation of virus development in a cell. Some evidence from *in vivo* studies supports the idea that the virus nucleic acid controls the shape of the inclusion body. STAIRS (1964) was able to select, from a virus preparation producing both oval and cuboidal inclusion bodies, as train of virus that produced only the cuboidal inclusion bodies.

No studies on the development of other nuclear polyhedrosis viruses in tissue culture have been reported. Also no reports exist of definitive studies of replication of cytoplasmic virus in tissue culture. Since these viruses contain RNA instead of DNA and since they are replicated in the cytoplasm instead of the nucleus their development is undoubtedly somewhat different from that described for the nuclear polyhedrosis viruses.

Of the noninclusion body viruses, only the iridescent viruses have been extensively studies in insects so it is not surprising that these viruses are the only members of this group so far studied in tissue culture. When BELLET (1965b) described the multiplication of the *Sericesthis* iridescent virus in the *Antheraea* cell line, he determined that formation of a stable virus-cell interaction was completed after incubation at 20° C for 10 hours. No decrease in the number of infected cells per culture occurred if the cells were washed with medium after 10 hours. BELLET, therefore, obtained infected cultures to determine the development of the virus by washing the cells after incubation at 20° C for one day. The first viral antigen demonstrable by fluorescent-antibody staining appeared on the fourth day; the first infective virus appeared 12 hours later. Thus this virus had an eclipse phase of from two to four days after inoculation.

Virus was produced and released continuously until the eighth day after inoculation; however, the amount of cell-associated virus was variable and only exceeded the amount of free virus on the sixth day. The mean latent period, that is the time required to produce an average of one infectious unit per infected cell, was calculated to be 5 days 14 hours. The total yield of infective virus was calculated as 495 infectious units per infected cell, of which approximately 358 per cell were released.

BELLET also showed that initiation of viral replication was a random process and thus was asynchronous. Replication began in some cells at two days after inoculation, and most infected cells had begun viral replication by six days. Replication began in all the foci of a multiple infected cell at about the same time, and the synthesis continued for about two days.

BELLET also studied the effect of infection on the synthesis of cell DNA as shown by incorporation of H^3-thymidine into the nucleus. In a culture, the percentage of cells (30%), which showed incorporation of H^3-thymidine into the nucleus was constant, whether only uninfected cells or both infected and uninfected cells were considered. When cells that showed cytoplasmic (viral) incorporation were considered, the percentage decreased by an amount significant at the 10% but not at the 5% level, and about 25% of the cells synthesized both viral DNA and nuclear DNA. Thus BELLET concluded that though some inhibition of nuclear DNA synthesis may have occurred, complete inhibition was not necessary for viral DNA synthesis.

Synopsis

Insect tissue cultures have long had much promise as a tool for the study of insect viruses. Improvements in the methodology within the last few years should permit some of this promise to be realized. The large body of literature now available on the cultivation of tissues from insects provides the investigator with a guide for the development of primary cultures that meet his needs, and and past work has established that all known types of insect viruses can be grown in such primary cultures. Continuous cultures of cells from two different insects has been achieved ,and at least two types of cytoplasmic viruses have been multiplied in the *Antheraea* cell line. Whenever it has been studied, the overall cycle of replication *in vitro* has been shown to be very similar to development *in vivo*.

Thus the necessary background for the study of the insect viruses *in vitro* is established; however, many aspects of virus replication have not received attention. Little work has been done to determine the mechanism or mechanisms by which the virus particle penetrates the cell membrane. Previous work has shown only that some specific conditions are required and the that process is accomplished in a few hours, even at low temperatures. Also extensive electron microscope studies on sections of tissue taken from intact insects have established the intracellular site of replication for many viruses. However, little is known about the dynamics of synthesis of the various components of the virus particles, and not much is known about the effect of virus infection on the normal synthesis patterns of the cell. In addition, the factors controlling the size and the shape of the inclusion bodies formed by many insect viruses have been of considerable interest to insect pathologists. Experience has shown that tissue cultures provide a suitable method for the study of the formation of these unique structures. These are but a few of the areas where more investigation is required and recent developments indicate that insect tissue culture should be of considerable value in these studies.

Cultivation of Plant Viruses in Insect Tissue Culture

All the work cited previously concerned the viruses of insects that are of economic importance in agriculture because they are destructive pests or because they produce material of economic value. Another large group of viruses, also of importance to agriculture, are those that cause diseases in plants. Since most such viruses are transmitted by insects, insect tissue culture should provide a useful tool for the study of the virus-vector relationship. However, few studies of this type have been made, again undoubtedly because methods of culturing tissue from the vectors have been lacking. Recently, much progress has been made in the culture of these tissues, (HIRUMI and MARAMOROSCH, 1964; MITSUHASHI and MARAMOROSCH, 1964; MITSUHASHI, 1965b).

The first evidence that the development of plant viruses could be studied in cultures of vector tissues was reported by MARAMOROSCH in 1956. Tissue was removed from aster leafhoppers that had been infected with aster yellows virus by feeding on infected aster plants. Part of this tissue was immediately homogenized, and the cell-free fluid was injected into virus-free insects. The remainder was cultured in hanging drop cultures for 10 days and then harvested, pooled, and homogenized. The cell-free fluid from the cultured material was injected into a second group of virus-free leafhoppers.

Leafhoppers injected with freshly collected material did not transmit the virus; insects injected with the cultured material did. This virus must undergo multiplication in an insect before it can be transmitted, and the evidence showed that the necessary multiplication could take place in tissue culture.

MITSUHASHI (1965c) published preliminary studies of the replication of the rice dwarf virus (RDV) in cultures of tissue from the embryos of the rice leafhopper, *Nephotettix cincticeps* (UHLER), a virus known to be transmitted transovarially from female adults to their progeny. To study this aspect of virus development, cultures of embryonic tissue from eggs produced by viruliferous females were established. Cell migration and multiplication occurred normally, and the epithelial cells formed cell sheets in a large area around the explants. However, the cells were more granular than normal, and considerable histolysis of the explants occurred simultaneously with the development of the cell sheets. After 27 days, one culture was sacrificed for examination with the electron microscope. RDV particles were found in almost all cells developed from the explant. The virus particles were arranged in long rows and surrounded by a sheath, an arrangement normally not found in the cells of intact insects.

MITSUHASHI also found that established cultures of embryonic leafhopper cells could be infected with RDV by inoculation with cell-free preparations from the fat body of viruliferous adults. Changes in cells occurred within 24 hours after inoculation. During the first three days, mitosis continued although the cells contained many granules. Granulation developed first around the nucleus but eventually filled the whole cytoplasm. As the granulation increased, the cells shrank, and the cell sheets were eventually destroyed. Most of the granules were approximately 30 mμ in diameter whereas normal RDV particles are 70 mμ in diameter. On the basis of their geometrical shapes MITSUHASHI considered them to be a form of the rice dwarf virus. Few normal size RDV particles were seen and infectivity tests were not completed so the exact nature of these particles is not known.

The infection of primary cultures of tissues from the leafhopper *Agallia constricta* VAN DUZEE infected with wound tumor virus (WTV) was studied by CHIU et al. (1966). They found that WTV produced no detectable cytopathic effects on the cultured cells. The continuously developing granulation observed by MITSUHASHI (1965c) with RDV did not occur. Also, the cell sheets did not deteriorate but continued to proliferate and to yield virus for many days. Two cultures were transplanted and continued to grow and form new cell sheets for as long as 64 days.

The presence of virus was demonstrated in the cultures by immunofluorescent staining and by determining the relative concentrations of infective WTV. These concentrations were determined from insects inoculated with the virus by using an experimentally derived regression curve to relate the concentration of virus injected into vectors to the WTV soluable antigen titer reached in the vectors 21 days later (REDDY and BLACK, 1966).

Viral antigens could be identified by immunofluorescence in a few cells at the third day after inoculation and were present for as long as 64 days. Viral antigen was found only in the cytoplasm and no similar staining with the fluorescein-conjugated antiserum occurred in the controls.

The infectivity studies showed that the relative virus concentrations of 10^8 in the inoculum decreased to 10^4 within 24 hours and then rose to a plateau level of approximately 10^6 after two days. In addition, seven serial passages were carried out. The concentration at the last passage was calculated to have been reduced to $10^{-2.5}$, however, determination of the actual level revealed a relative virus concentration of $10^{5.4}$ to $10^{6.1}$. Thus, multiplication of the virus during the series of passages was established.

Although only a few reports exist of the growth of plant viruses in cultures of tissue from their vectors, the results are sufficient to establish the usefulness of such a system to obtain important information concerning the replication of the plant viruses during the insect phase of their development cycle. The demonstration by MITSUHASHI (1965c) that virus can be cultured in the tissue of embryos from viruliferous females is of considerable importance, not only where plant viruses are involved, but wherever transovarian transmission of viruses in insects is suspected.

Cultivation of Arboviruses in Arthropod Tissue Culture

The arboviruses comprise another group of infectious agents for which the insects serve as both a host and a vector. Establishment of this relationship depends upon satisfying four criteria: detection of the range of naturally infected species, determination of the minimal virus titer necessary to infect, demonstration of virus multiplication in the insect, and, finally, establishment of the vector's capacity to transmit the virus by bite (CASALS and REEVES, 1965). The second and third of these four criteria are amendable to study in cultures of tissue from vectors. Also where a mechanism for prolonged survival in the tissue of the arthropod host occurs, or is suspected, tissue cultures may also be of value in elucidating this mechanism. Although some 200 viruses have been tentatively classified as Arboviruses and four or possibly five, families of blood-sucking arthropods are implicated as natural vectors, little use has been made of insect tissue cultures to study this virus-host relationship.

The first attempt to grow any arboviruses in insect tissue culture was reported by TRAGER in 1938. He maintained fragments of tissues from the mosquito, *Aedes aegypti* for two to three weeks depending upon the tissue. The virus used was a strain of western equine encephalomyelitis virus maintained by mouse brain passage; the cultures were inoculated with five per cent brain suspension made up in the nutrient solution and incubated at 28° C.

In the controls, which consisted of the brain suspension and plasma but no tissue, some virus was present after four days but not after seven days. Virus survived for seven days in only three of the eight cultures of adult ovaries. In the second passage of these three cultures, the virus was present in only one, though titration showed a 100 fold increase. In cultures of the adult gut, the virus survived with only slight loss in titer, but a second passage was not successful. In cultures of larval gut, a second passage of the virus was successful, but a third passage failed. Thus neither of these gut tissue nor tissue from the thoracic body wall survived well in culture, and, as might be expected, none supported much replication of the virus. These experiments indicate that the tissue exerted a protective effect but supported little virus replication.

In a second experiment, TRAGER passed the virus through alternating cultures of larval thoracic tissue and pupal head tissue for four successive passages. At the end of the fourth passage, the virus titered 10^5 times higher than what could be accounted for on the basis of the original inoculum, an increase in each passage of approximately 100-fold.

PELEG and TRAGER (1963) reported similar results with the Egypt 101 West Nile virus in primary cultures of imaginal discs of *Aedes aegypti* larvae. The virus was passed in mice, and infected mouse brain suspensions were used as inoculum. The log of LD_{50} in the inoculum was approximately 4.5; by the second day after inoculation, the titer had decreased to 3.4, but then an increase in the virus level began. By the fifth day, the virus level, log $LD_{50} = 5.0$, exceeded that of the original inoculum and the peak concentration, log $LD_{50} = 6.5$, was reached on the tenth day. Thus, the virus level increased 100-fold over the inoculum and 1,000 fold over the lowest level reached.

A total of four passages were made, each after 10 days' incubation of the previous culture. The virus recovered per culture from the fourth passage was equal to that of the inoculum used in the first. Since the mechanics of passage resulted in a 1.6×10^{-5} dilution of the original virus, multiplication occurred.

The first studies on arboviruses in cultures of tick tissue were reported in 1960 when REHACEK and PESEK published work on the replication of eastern equine encephalomyelitis (EEE) in primary tick cultures. EEE virus multiplied approximately 100-fold in cultures of fat body and hypodermis tissue from the tick, *Dermacentor pictus*, but cultures of tissue from the salivary gland, gut, ovary, and malpighian tubes showed a decrease in virus titer of from one to two logs over the same period of time. Large inoculum, 10^4 $TCID_{50}$, in cultures of all tissues did not initiate multiplication; instead, the tissue provided only a protective effect that prevented the inactivation of the virus that occurred in the controls. When the tissues were inoculated with either $10^{2.5}$ or $10^{0.6}$ $TCID_{50}$, virus multplication occurred. Similar results were obtained with EEE virus in cultures of the fat body and hypodermis from *Ixodes ricinus* (L.). However, these same tissues from a third tick, *Dermacentor marginatus*, only protected the virus from inactivation.

REHACEK (1963) found that cultures of tissue from both *Hyalomma asiaticum* SCHULZE and SCHLATTKE, and *D. marginatus* supported multiplication of TE virus, strain Hypr, even though only *D. marginatus* is a known vector of this virus. In cultures of tissue from *H. asiaticum*, the level of virus increased by 1 to 5 logs over a period of eight days as measured by intracerebral inoculation of mice. Again, the most clear cut demonstration of virus multiplication was obtained with low concentrations of virus. High concentrations (10^4 LD_{50}) resulted in a very short period of multiplication and a relatively small increase in the amount of virus. The peak level was reached in 4 to 6 days and was followed by a decline. When a low concentration was used in the inoculum the virus titer continued to increase for eight days, the entire duration of the experiments. What would have happened in a longer experiment is not known. REHACEK found that the control primary cultures normally survived for more than 20 days without changes of medium, but he did not describe the fate of infected cultures over a similar period of time.

As noted, the best multiplication of TE virus was obtained when very low concentrations of virus were used as an inoculum, indicating that the primary tick cultures might be a sensitive system for detecting low concentrations of this virus. To test this assumption, REHACEK and KOZUCH (1964) compared the susceptibility of cultures of tissue from ticks with chick embryo cell cultures. Cultures of both cells were inoculated with low concentrations of the Hypr strain of TE virus carried in mice. After five days, subcultures were made to chick embryo cell cultures, and the presence of TE virus was demonstrated by a method based on interference with WEE virus. When inocula containing 30 and 3 $TCID_{50}$ were used, the percentage of infected chick embryo cell and tick cell cultures was about equal. However, when the inocula were reduced to 0.3 and 0.03 $TCID_{50}$ the respective percentages of infected tick cell cultures were 76 and 18, whereas, that of chick embryo cell cultures were only 34 and 3, respectively. Thus, the tick culture was a more reliable indicator of low concentrations of the TE virus.

REHACEK (1965) made a very extensive survey to try to determine the range of viruses that would multiply in primary cultures of tissues from *Hyalomma dromedarii* KOCH. Strains of both group A and B arboviruses and six viruses

Table 1. *Survey of viruses tested in H. dromedarii tissue culture*[a]

Virus	Strain	Titrated in[b]	Maximum titer[c]
ARBO A			
EEE	28 N 8	Mice, CEC	5
WEE	—	Mice, CEC	4
Semliki	—	CEC	5
Sindbis	AR 339	CEC	5
ARBO B			
JBE	P 3	Mice	2
West Nile	Egypt 101	Mice	3
St. Louis	—	Mice	2
Yellow fever	Asibi	Mice	1
ETBE	Hyper	Mice	5
KFD	W 377	Mice	3
Langat	TP 21	Mice	2
Louping-ill	Li 59	Mice	5
OHF	—	Mice	2.5
Powassan	—	Mice	2.5
RSSE	Sofin	CEC	5
NON-ARBO			
LCM	3b	Mice	4
EMC	—	MEC	—
Polio	MEF 1	HELA	—
VS	—	MEC	—
Vaccinia	—	HELA	—
NDV	Herfordshire	CEC	—
Pseudorabies	BUK	CEC	—
Pseudorabies	CVHD	CEC	—

[a] From REHACEK, 1965.
[b] CEC = Chick embryo cell cultures. MEC = Mouse embryo cell cultures.
[c] Max. titer reached in tick tissue cultures (log $TCID_{50}$ or log mouse LD_{50}).

that are not in the arbovirus group were tested (Table 1). Since all tests were single passages, the protective effect of the tissues could not be measured. Thus determination of actual multiplication depends entirely on the comparison of the final titer with that of the inoculum (shown only in line graphs). When this difference was substantial, as in the case of Semliki forest virus, Sindbis, or ETBE, multiplication is implied. Often, the increase was slight e.g., Japanese B encephalitis, and St. Louis encephalitis, and the protective effect of the tissue may simply have permitted survival of the inocula with little or no multiplication. The only virus not a member of the arbovirus group that gave evidence of multiplication in the primary tick cell culture was LCM.

The study of these viruses in arthropod tissue cultures has, of course, been limited by the difficulty in establishing and maintaining sufficient cultures, but only one attempt to use either of the two cell lines developed by GRACE has been reported (SUITOR, 1966). The virus used was a strain of Japanese encephalitis virus, and the cultures were grown in a medium containing 10% *A. pernyi* hemolymph. SUITOR found this level of hemolymph permitted virus replication, though, 2% hemolymph, the amount required for growth of the cells, did not. Cultures were incubated at 28° C for 35 days without changing medium, a procedure that was undoubtedly detrimental to the cells.

Total virus levels, intracellular plus free virus, were determined by titration in mice and by PORTERFIELD'S plaque inhibition test using the MK-2 cell culture system. These tests showed more than a 10-fold initial decrease in the level of virus followed by a sharp increase beginning after the seventh day and reaching a peak after 28 days. The peak was 10—100 times greater than the inoculum and 100 to 1000-fold higher than the concentration present after the loss of infectivity the first week.

SUITOR'S results are the first evidence of any substantial multiplication of an arbovirus in arthropod tissue cultures from other than mosquito or ticks. The availability of the two GRACE lines should provide much more convenient insect systems for the study of this group of viruses. Also from the evidence reported thus far, the insect cell systems are susceptible to many arboviruses. Thus, as suggested by REHACEK they may provide an excellent tool for detection of the vectors of these viruses. With the improvement in the insect cell culture methods, studies of the virus-vector relationship comparable to those of the virus-host relationship should become feasible.

References

AIZAWA, K.: Mode of multiplication of silkworm nuclear polyhedrosis virus. II. Multiplication of the virus in the early period of the LD_{50}-time curve. J. Insect Pathol. **1**, 67—74 (1959).

—, and C. VAGO: Culture *in vitro* de cellules separées de tissue d'insects. Compt. Rend. **249**, 928—930 (1959a).

— — Sur l'infection a Borrelinavirus en culture de tissue d'insectes. Ann. Inst. Pasteur **96**, 455—460 (1959b).

BELLETT, A. J. D.: The multiplication of *Sericesthis* iridescent virus cell cultures from *Antheraea eucalypti* SCOTT. II. An *in vitro* assay for the virus. Virology **26**, 127—131 (1965a).

— The multiplication of *Sericesthis* iridescent virus in cell cultures from *Antheraea eucalypti* SCOTT. III. Quantitative experiments. Virology **26**, 132—141 (1965b).

—, and E. H. MERCER: The multiplication of *Sericesthis* iridescent virus in cell cultures from *Atheraea eucalypti* SCOTT. I. Qualitative Experiments. Virology **24**, 645—653 (1964).

CASALS, J., and W. C. REEVES: The arboviruses. In: Viral and rickettsial infections of man, 4 ed. (F. L. HORSFALL and I. TAMM, eds.). Philadelphia: J. B. Lippincott Co. 1965.

CHIU, R., D. V. R. REDDY, and L. M. BLACK: Inoculation and infection of leafhopper tissue cultures with a plant virus. Virology **30**, 562—566 (1966).

FAULKNER, P., and J. L. VAUGHN: Insect tissue cultures for the study of silkworm nuclear polyhedrosis virus. Proc. XII Intern. Congr. Entomol. London 1964, p. 717—720. 1965.

GAW, S., N. T. LIU, and T. U. ZIA: Tissue culture methods for cultivation of virus grasserie. Acta virol. **3**, Suppl., 55—60 (1959).

GRACE, T. D. C.: Induction of polyhedral bodies in ovarian tissues of the tussock moth *in vitro*. Science **128**, 249—250 (1958).

GRACE, T. D. C.: Tissue culture for arthropod viruses. Trans. N.Y. Acad. Sci. **21**, 237—241 (1959).
— Establishment of four strains of cells from insect tissues grown *in vitro*. Nature (Lond.) **195**, 788—789 (1962a).
— The development of cytoplasmic polyhedrosis in insect cells grown *in vitro*. Virology **18**, 33—42 (1962b).
— Establishment of a line of mosquito (*Aedes aegypti* L.) cells grown *in vitro*. Nature (Lond.) **211**, 366—367 (1966).
HEITOR, F.: Étude de l'infectiosité de l'hemolymphe en vue de l'introduction de virus de polyedries en culture de tissue d'insectes. Ann. Epiphyties **14**, 213—215 (1963).
HIRUMI, H., and K. MARAMOROSCH: The *in vitro* cultivation of embryonic leafhopper tissues. Exp. Cell Res. **36**, 625—631 (1964).
HORIKAWA, M., and A. S. FOX: Culture of embryonic cells of *Drosophila melanogaster in vitro*. Science **145**, 1437—1439 (1964).
HUGER, A.: Granuloses of insects. In: Insect pathology, vol. I. (E. A. STEINHAUS, ed.). New York: Academic Press 1963.
JONES, B. M.: The cultivation of insect cells and tissues. Biol. Rev. **37**, 512—536 (1962).
—, and I. CUNNINGHAM: Growth by cell division in insect tissue culture. Exp. Cell Res. **23**, 386—401 (1961).
LURIA, S. E.: General virology 1st ed. New York: John Wiley & Sons 1953.
MARAMOROSCH, K.: Multiplication of aster yellows virus in *in vitro* preparations of insect tissues. Virology **2**, 369—376 (1956).
MARTIGNONI, M. E.: Insect tissue culture: A tool for the physiologist. Insect Physiol. Proc. 23rd Biol. Colloq. Oregon State Univ., vol. **23**, p. 89—110 (1962).
—, and R. J. SCALLION: Preparation and uses of insect hemocyte monolayers *in vitro*. Biol. Bull. **121**, 507—520 (1961a).
— — Multiplication *in vitro* of a nuclear polyhedrosis virus in insect amoebocytes. Nature (Lond.) **190**, 1133—1134 (1961b).
— E. M. ZITCER, and R. P. WAGNER: Preparation of cell suspensions from insect tissues for *in vitro* cultivation. Science **128**, 360—361 (1958).
MITSUHASHI, J.: Tissue culture of the rice stem borer, *Chilo suppressalis* WALKER (Lepidoptera: Pyralidae) I. Cell migration from the explanted tissues of diapausing larvae. Japan. J. appl. Ent. Zool. **9**, 217—224 (1965a).
— *In vitro* cultivation of the embryonic tissues of the green rice leafhopper, *Nephotettix cincticeps* UHLER (Homoptera: Cicadellidae) Japan. J. appl. Ent. Zool. **9**, 107—114 (1965b).
— Preliminary report on the plant virus multiplication in the leafhopper vector cells grown *in vitro*. Japan. J. appl. Ent. Zool. **9**, 137—141 (1965c).
— Tissue culture of the rice stem borer, *Chilo suppressalis* WALKER (Lepidoptera: Pyralidae) II. Morphology and *in vitro* cultivation of hemocytes. Appl. Ent. Zool. **1**, 5—20 (1966a).
— Multiplication of *Chilo* iridescent virus in the *Chilo suppressalis* tissues cultivated *in vitro* (Lepidoptera: Pyralidae). Appl. Ent. Zool. **1**, 199—201 (1966b).
—, and K. MARAMOROSCH: Leafhopper tissue culture: embryonic, nymphal, and imaginal tissues from aseptic insects. Contr. Boyce Thompson Inst. **22**, 435—460 (1964).
PELEG, J.: Growth of *Aedes aegypti* embryonic cells and tissues *in vitro*. Experientia (Basel) **22**, 555—558 (1966).
—, and W. TRAGER: Cultivation of insect tissues *in vitro* and their application to the study of arthropod-borne viruses. Amer. J. trop. Med. Hyg. **12**, 820—824 (1963).
REDDY, D. V. R., and L. M. BLACK: Production of wound-tumor virus and wound-tumor soluble antigen in the insect vector. Virology **30**, 551—561 (1966).

REHACEK, J.: Propagation of tick-borne encephalitis (T.E.) virus in tick tissue cultures. Ann. Epiphyties **14**, 199—204 (1963).
— Cultivation of different viruses in tick tissue cultures. Acta. virol. **9**, 332—337 (1965).
—, and O. KOZUCH: Comparison of the susceptibility of primary tick and chick embryo cell cultures to small amounts of tick-borne encephalitis virus. Acta virol. **8**, 470—471 (1964).
—, and J. PESEK: Propagation of eastern equine encephalomyelitis (EEE) virus in surviving tick tissues. Acta virol. **4**, 241—245 (1960).
SEN GUPTA, K.: Multiplication of a nuclear polyhedral virus in the fat bodies and blood cells of *Galleria mellonella* L. cultured *in vitro*. Indian J. exp. Biol. **1**, 222—224 (1963).
—, Studies on the specificity of haemolymph from different insects for the culture of tissues from other insects. Proc. ind. Acad. Sci. **59**, 103—109 (1964).
SMITH, K. M.: Insect virology. New York: Academic Press 1967.
STAIRS, G. R.: Selection of a strain of insect granulosis virus producing only cubic inclusion bodies. Virology **24**, 520—521 (1964).
SUITOR jr., E. C.: Growth of Japanese encephalitis virus in Grace's continuous line of moth cells. Virology **30**, 143—144 (1966).
TING, K. Y., and M. A. BROOKS: Sodium: potassium ratios in insect cell culture and the growth of cockroach cells (Blattariae: Blattidae). Ann. Entomol. Soc. Amer. **58**, 197—202 (1965).
TRAGER, W.: Cultivation of the virus of grasserie in silkworm tissue cultures. J. exp. Med. **8**, 501—514 (1935).
— Multiplication of the virus of the equine encephalomyelitis in surviving mosquito tissues. Amer. J. trop. Med. **18**, 387—393 (1938).
VAGO, C.: Cultures de tissue d'insectes a l'aide de serum de mammifères. Entomophaga **7**, 175—179 (1962).
—, and M. BERGOIN: Developpement des virus a crops d'inclusion du Lepidoptere *Lymantria dispar* en cultures cellularies. Entomophaga **8**, 253—261 (1963).
—, and S. CHASTANG: Obtention de lignees cellulaires en culture de tissue d'invertébres. Experientia (Basel) **14**, 110—113 (1958).
— — Culture de Borrelinavirus dans les organes d'insectes en survie. Compt. Rend. **251**, 903—905 (1960).
—, and O. CROISSANT: Microscopie electronique de cultures de tissue d'insectes saines et infecties de virus. Ann. Epiphyties **14**, 43—51 (1962).
— J. QUIOT, and J. LUCIANI: Infection de culture de tissue de lepidopteres avec le virus purifie de la densonucleose. Compt. Rend. **263**, 799—800 (1966).
VAUGHN, J. L.: A tissue culture method for studying insect viruses. Bact. Proc. 131 (1963).
— Chromatographic evidence of differences between the virus in the hemolymph and in the polyhedra from diseased *Bombyx mori*. J. Invert. Pathol. **7**, 524—525 (1965).
—, and P. FAULKNER: Susceptibility of an insect tissue culture to infection by virus preparations of the nuclear polyhedrosis of the silkworm (*Bombyx mori* L.). Virology **20**, 484—489 (1963).
WYATT, S. S.: Culture *in vitro* of tissue from the silkworm *Bombyx mori* L. J. gen. Physiol. **39**, 841—852 (1956).
YUNKER, C. E., J. L. VAUGHN, and J. CORY: Adaptation of an insect cell line (Grace's *Antheraea* Cells) to medium free of insect hemolymph. Science **155**, 1565—1566 (1967).

Viruses — Living Insecticides

CARLO M. IGNOFFO*

With 20 Figures

Introduction

Excluding himself, one of man's greatest competitors for food, clothing, and shelter are insects. Man probably first fought the insect in the fertile crescent when he plucked a handful of wild wheat seed, planted it, nursed it, and then harvested it for food. He's still fighting the same insects on the same crops.

Sophisticated, efficient chemical insecticides were developed during World War II and widely used after the war until recently. We are now concerned that chemicals are more efficient than they should be. Not only do they kill the *bad guys*, but also the *good guys* — domestic animals, wild game, and even man. No sensible entomologist advocates outlawing chemical insecticides. However, because of insect resistance, toxicity, environmental pollution, residues on food, and legal restrictions, other more selective methods of insect control should and must be developed. Selective insect control — in this specific case — insect control using living biological, viral-insecticides is one of the available approaches.

Insects have the same diseases we have. They are infected by protozoa, bacteria, viruses, fungi, and rickettsiae. Over 1,000 insect diseases are described. Viruses represent approximately 25 percent of this total.

Virus diseases of insects have been known since the Civil War. However, it is only during the last few years that any intensive research has been conducted on their possible use for controlling arthropod pests. This work, initiated after World War II, was kindled when it became obvious that problems were developing with the continued use of chemical insecticides.

PASTEUR said, "There are not two different kinds of science; there is science and there are applications of science. Two types constantly interplay and one type cannot progress far without contact with the other." (DUBOS, 1960). How vividly this applies to arthropod viruses becomes evident as I evaluate living viral insecticides.

Availability of Living Viral Insecticides

Types of Arthropod Viruses

Nearly 300 viruses are available for use against specific agricultural pests and more are described each year. As a minimal estimate, 12—15 "new virus

* International Minerals and Chemical Corp. Libertyville, Illinois, 60048 U.S.A.

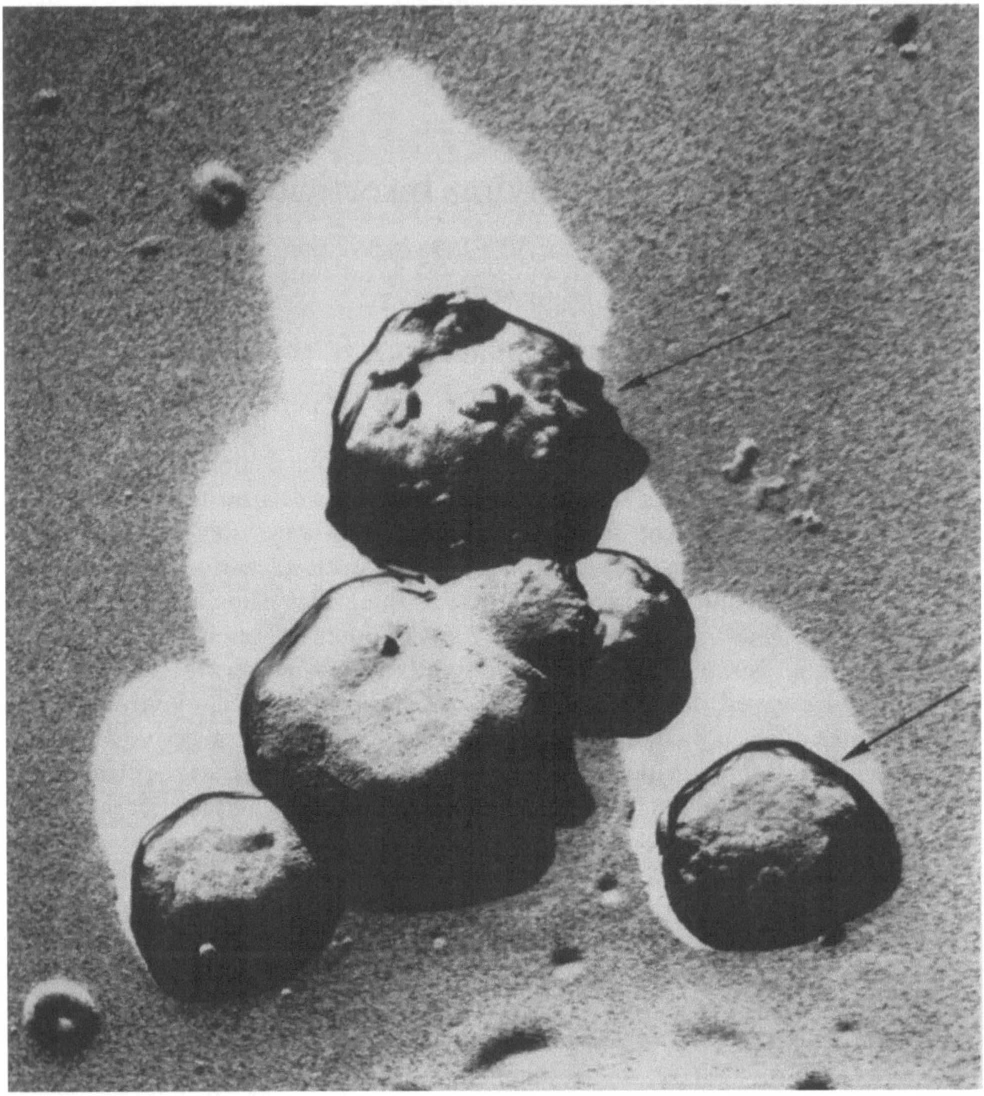

Fig. 1. Nuclear polyhedrosis virus of corn earworm; carbon replica of inclusion body (arrow); ×23,850. (Photo by J. R. ADAMS)

species" are described each year. Ninety-five percent of the arthropod viruses have a unique, proteinaceous, crystalline-like structure called an inclusion body (Fig. 1). Infective viral particles or virions (Fig. 2), occluded in the protein matrix of these inclusions, account for about 1 percent of the total protein. Inclusion bodies are detected with ordinary light microscopy and this, in effect, accounts for their early description and reportedly largest representation among the arthropods. Inclusion viruses were first independently described by MALESTRI (1856) and CORNALIA (1856) from diseased silkworms. Non-inclusion viruses (those which do not have a crystalline-like body) make up the remaining 5 percent of the described viruses (Fig. 3). The first of the

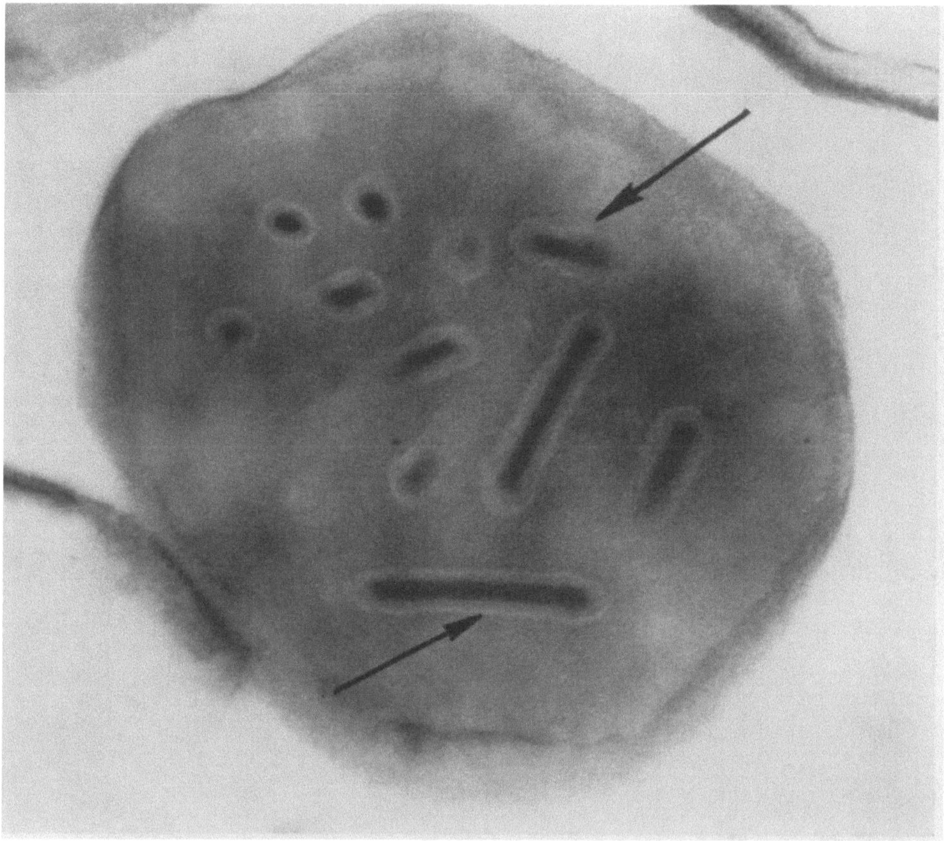

Fig. 2. Nuclear polyhedrosis virus of corn earworm; sectioned inclusion body with occluded-rod-shaped virions (arrows); ×112,000. (Photo by J. R. ADAMS)

Fig. 3. Non-inclusion virus of citrus red mite; shadowed virions (arrow); ×76,300. (Photo by J. R. ADAMS)

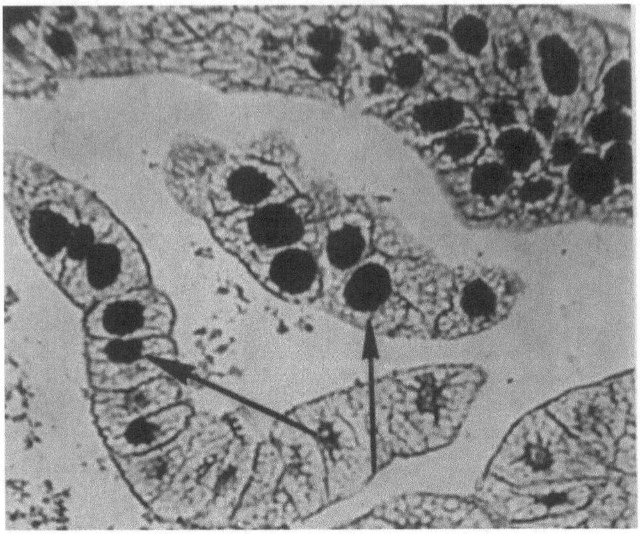

Fig. 4. Nuclear polyhedrosis virus of corn earworm nucleus of fat cells packed with inclusion bodies (arrow); ×500. (Photo by J. HAMM)

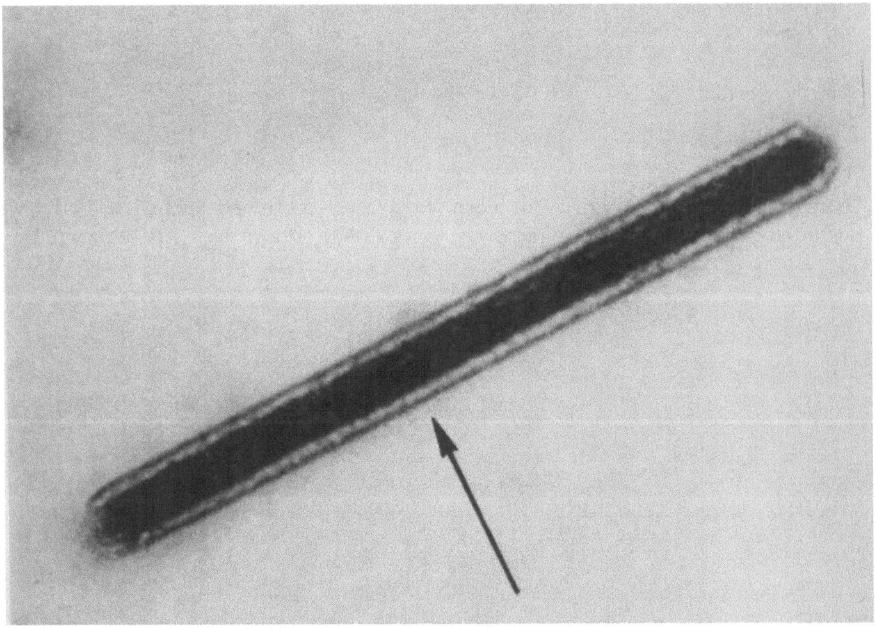

Fig. 5. Nuclear polyhedrosis virus of corn earworm; virion (arrow) ×220,000. (Photo by J. R. ADAMS)

non-inclusion type viruses was described less than 15 years ago (WASSER, 1952). Virions of the non-inclusion viruses are morphologically similar to plant and animal viruses and require electron microscopy for their detection. Undoubtedly, more non-inclusion viruses will be described from broader spectrum of invertebrates as the electron microscope becomes a regular tool of insect virologists.

Presence or absence of an inclusion body is a major criterion when classifying arthropod viruses (Table 1). Further differentiation of inclusion viruses is based on the inclusion body shape and site of viral replication. Inclusion bodies with sharp, distinct sides, forming regular or irregular polyhedrons are called polyhedroses. Eighty percent of all arthropod virus diseases are polyhedroses. When inclusion bodies are formed in the cytoplasm, the diseases are called cytoplasmic polyhedroses. Conversely, nuclear polyhedroses are diseases where inclusion bodies develop in the nuclei of infected cells. The nuclear and cytoplasmic polyhedroses represent about 44 and 36 percent, respectively, of all described viral diseases.

Fig. 6. Dead cabbage looper larvae, *Trichoplusia ni*, killed by its nuclear polyhedrosis virus; ×2

All nuclear polyhedroses (122 have been isolated from arthropods) are DNA viruses (Table 2). Inclusions are formed in the nuclei of blood, epidermal, adipose, and tracheal cells (Fig. 4). Polyhedral inclusion bodies (PIB) average 200 to 2,000 millimicrons in diameter, but some reach diameters of 15,000 millimicrons. Approximately 10 to 100 virions are imbedded, singly or in bundles, within each inclusion body. The virions (Fig. 5), which overtly resemble the rod-shaped, helically-symmetrical, tobacco mosaic virus (TMV), average 400×80 millimicrons. Based upon a TMV model, each virion would contain approximately 2,500 capsomeres.

It is impossible to describe symptoms which apply to all nuclear polyhedroses. About 2 to 7 days after infection, the usually clear hemolymph becomes increasingly cloudy. One to two days later, the caterpillar becomes flaccid, dies, and the skin ruptures releasing billions of inclusion bodies. Twenty to 30 percent of the dry weight of caterpillars is inclusion bodies. In nature, dead, diseased caterpillars are characteristically posed at plant terminals hanging by their heels (Fig. 6).

The cytoplasmic polyhedroses only develop in cytoplasm of epithelial cells of the digestive tract, i.e., midgut (Table 2, Fig. 7). Approximately 100 viruses of this type were described from arthropods. All are RNA viruses. Some inclusions are regular dodecahedrons or icosahedrons (Fig. 8). They average 500 to 2,500 millimicrons in diameter, and contain an estimated 100 to 1,000 virions. Individual polyhedrons 25,000 millimicrons in diameter were also observed. Occluded virions (Fig. 9) are sub-spherical in shape (some are icosahedrons) and morphologically resemble small DNA-RNA virus of verte-

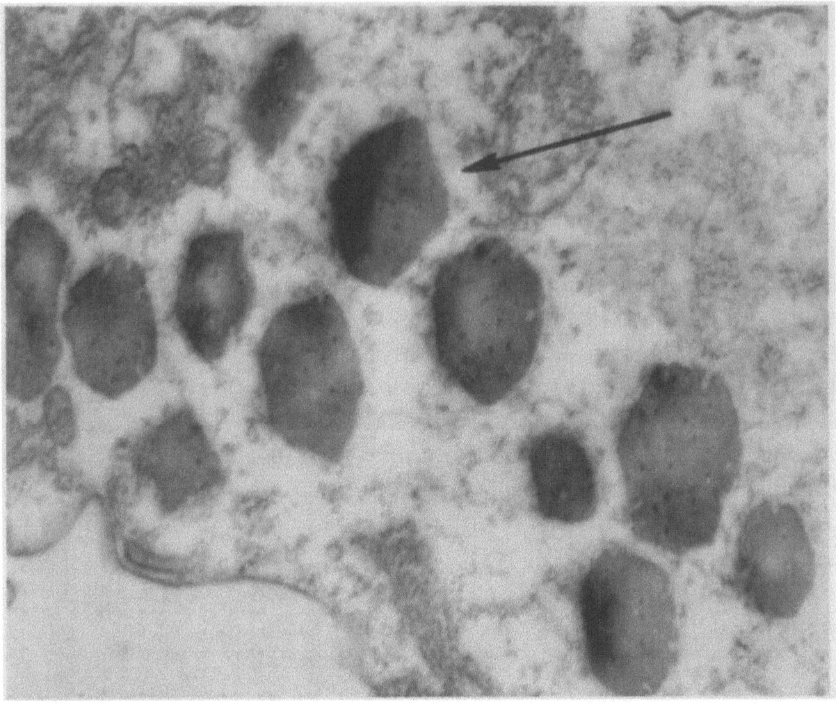

Fig. 7. Cytoplasmic polyhedrosis virus of pink bollworms sectioned inclusion bodies in cytoplasm of midgut epithelial cell (arrow); ×9,500. (Photo by J. R. ADAMS)

Table 1. *Classification of arthropod viruses*

Phylum	Inclusion body presence or absence	Shape of inclusion body	Type of disease	Primary site of viral replication: cell	Primary site of viral replication: tissue	Generic name
Arthropoda	Presence	Polyhedron	Polyhedroses	Nucleus	Mesoderm, ectoderm, endoderm	*Borrelinavirus*
						Birdiavirus
				Cytoplasm	Midgut epithelium	*Smithiavirus*
		Ovoid	Granuloses	Nucleus	Adipose, epidermus	*Steinhausiavirus*
				Cytoplasm and nucleus	Adipose, epidermus	*Bergoldiavirus*
		Irregular	Polymorphic inclusion viroses	Cytoplasm	Adipose, blood	*Paillotiavirus*
		Fusiform		Cytoplasm	Adipose	*Vagoiavirus*
	Absence	None	Non-inclusion viroses	Cytoplasm	Adipose-RNA	*Moratorvirus*
					Adipose-DNA	*Pseudomoratovirus*

Table 2. *Characteristics of the major arthropod viruses*

Virus disease	Replication site	Type genetic material	Inclusion body		Virion			Estimated virions per inclusion
			shape	size (mμ)	shape	symmetry	size (mμ)	
Nuclear polyhedroses	Nucleus of mesodermal, endodermal, and ectodermal cells	DNA	polyhedron	200—2,000	rod	helical	400×80	10—100
Cytoplasmic polyhedroses	Cytoplasm of midgut epithelial cells	RNA	polyhedron	500—2,500	spherical	cubical	30 to 80	100—1,000
Granuloses	Nucleus and cytoplasm of adipose, tracheal, and epidermal cells	DNA	ovoid	200×400	rod	helical	350×50	1—2
Polymorphic viroses	Cytoplasm of adipose cells, hemocytes	—	irregular	5,000—15,000	—	—	—	—
	Cytoplasm of adipose cells, hemocytes	—	fusiform	6,000—13,000	oval	—	400×250	10—100
Non-inclusion viroses	Cytoplasm of adipose cells	DNA RNA	none	none	spherical	cubical	35 to 150	
		DNA	none	none	rod	helical	200×10	

Fig. 8. Cytoplasmic polyhedrosis virus of pink bollworm; (arrow) carbon replica of inclusion bodies; ×15,000. (Photo by J. R. ADAMS)

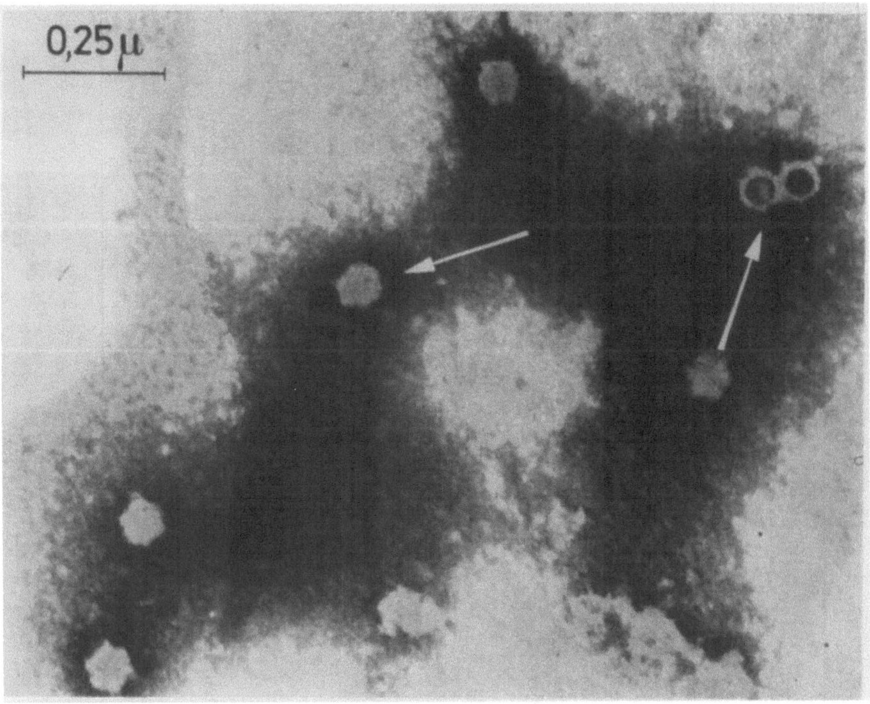

Fig. 9. Cytoplasmic polyhedrosis virus of pink bollworm; virions negatively stained (arrows); ×80,000. (Photo by J. R. ADAMS)

brates (warts, adenovirus, herpes, mumps). Arthropod virions averaging 30 to 80 millimicrons in diameter may contain 12 to 800 capsomeres. Caterpillars infected with cytoplasmic polyhedroses are easily distinguished from those infected with nuclear polyhedroses, especially in later stages of the disease. Diseased caterpillars develop slower than their healthy litter mates. In later stages, the midgut shows through the skin as a distinct, chalk-white patch

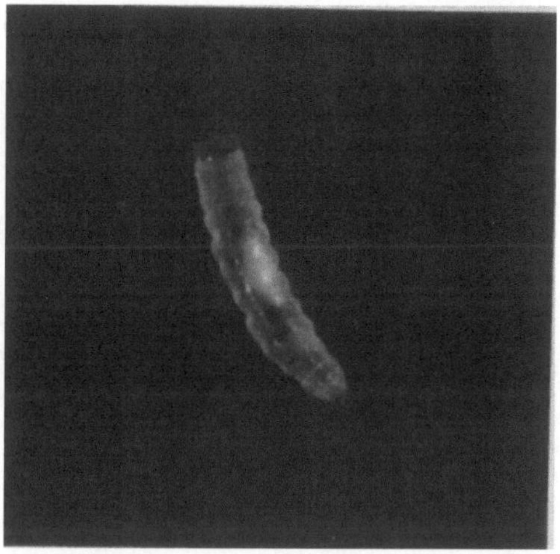

Fig. 10. Diseased pink bollworm larvae, *Pectinophora gossypiella*, cytoplasmic polyhedrosis virus; ×3,2

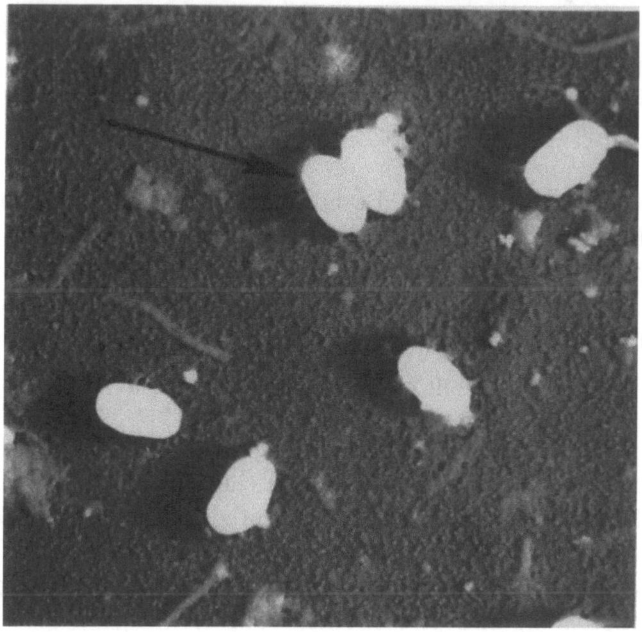

Fig. 11. Granulosis virus of codling moth, *Carpocapsa pomonella*; shadowed inclusion body (arrow); ×31,000. (Photo by R. LEUTENEGGER, courtesy L. FALCON, Y. TANADA)

(Fig. 10). This color is due to an accumulation of inclusion bodies. The skin of diseased caterpillars does not rupture as it does in the nuclear polyhedroses.

Another group, the granuloses, produce ovo-cylindrical, capsule-like inclusions (Table 2). Granuloses represent about 14 percent of all arthropod viruses and replicate in either nuclei or cytoplasm of adipose, tracheal, matrix, and epidermal cells. All granuloses are DNA viruses. Inclusion bodies are ovoid in shape and measure 200×400 millimicrons (Fig. 11). One, sometimes two, rod-shaped virions (Fig. 12) averaging 50×350 millimicrons, are

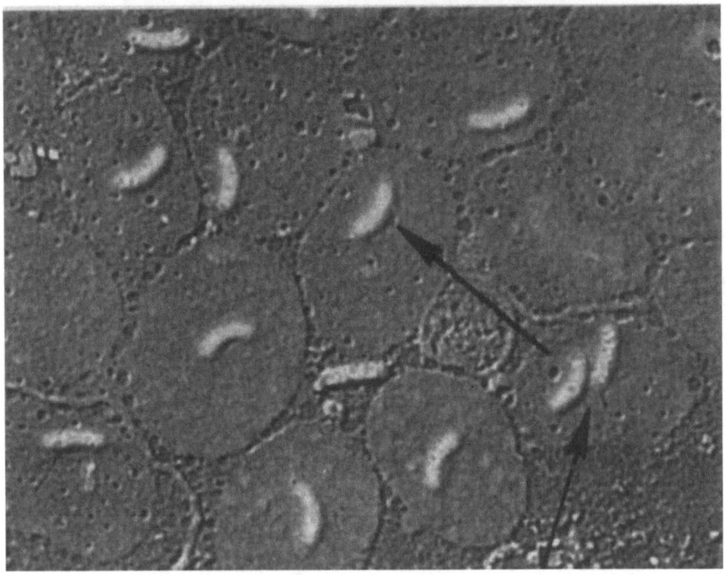

Fig. 12. Granulosis virus of codling moth. Partially dissolved inclusion body showing released, rod-shaped virion (arrow); ×31,000. (Photo by R. LEUTENEGGER, courtesy L. FALCON, Y. TANADA)

embedded in each inclusion body. Based upon a TMV model, each rod (Fig. 13) contains about 2,000 capsomeres. The symptomatic progression and termination of granulosis infections resemble nuclear polyhedroses infections. Body contents are completely liquified and the caterpillar's skin usually ruptures releasing billions of inclusions.

The last type of inclusion virus is represented by only one disease, i.e., polymorphic inclusion virus of the cabbageworm, *Pieris rapae* (PAILLOT, 1924, 1926). Large (5,000 to 15,000 millimicrons), irregular shaped inclusions form in the cytoplasm of infected fat and blood cells (Table 2). No crystalline-like matrix was described nor were virions detected in these inclusions. Recently, a new type of inclusion body, which might be included in this group, was described by WEISER (1965) and WEISER and VAGO (1966) from a beetle, *Melolontha melolontha*, and two moths, *Opherophtera brumata* and *Acrobasi zelleri*. The inclusions are ovoid-spindle-fusiform in shape and 6×13 microns in size. The virus replicates only in the cytoplasm of fat cells (Table 2). Large, oval (250×400 millimicrons) virions are embedded in the inclusion bodies. Based upon published photographs, an estimated 10 to 100 virions are embedded in each inclusion body. Disease symptoms are different from that of polyhedroses, granuloses, and the non-inclusion viroses. After death, the insect becomes rigid and hard. A new genus, *Vagoiavirus*, was established for this virus (WEISER, 1965).

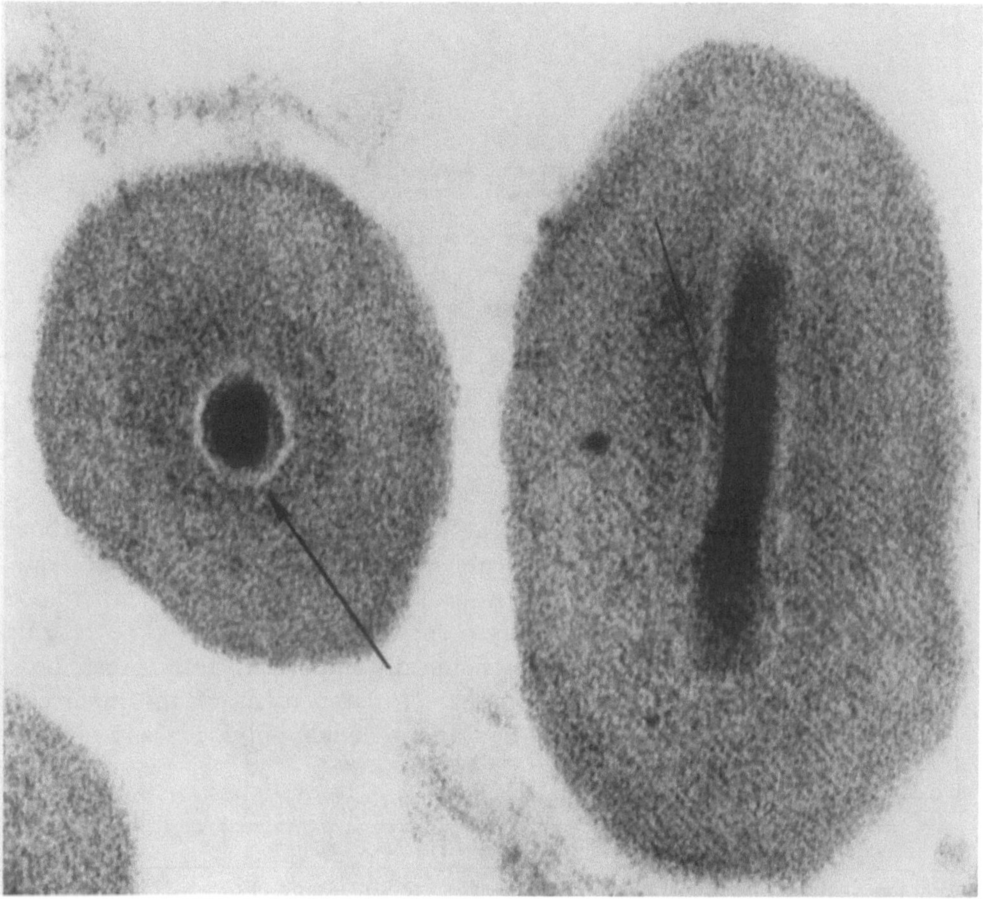

Fig. 13. Granulosis virus of red-banded leaf roller, *Agryrotaenia velutinana*, sectioned inclusion body showing occluded virion (arrow); ×118,500. (Photo by J. R. ADAMS)

Most of the non-inclusion virions are subspherical to spherical in shape (Figs. 3, 14) and average 35 to 150 millimicrons in diameter (Table 2). Rod-shaped particles (virions) were recently detected in the fruit fly, *Drosophila spp.* (BERKALOFF, et al., 1965), and the alkali bee, *Nomia melanderi* (THOMSON and SMIRK, 1966). Based upon known plant and vertebrate viruses, approximately 812 capsomeres are found in virions with diameters of 120—130 millimicrons. Both RNA and DNA viruses are described. Replication is generally associated with albuminoid granules located in the cytoplasm of adipose cells and with the endoplasmic reticulum of adipose cells.

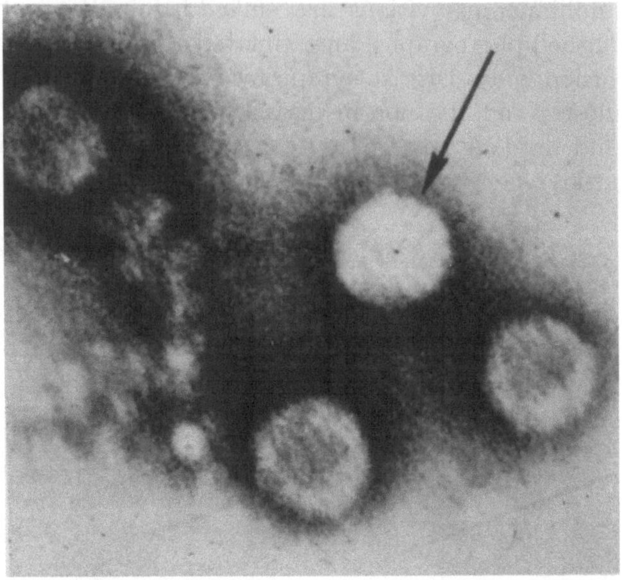

Fig. 14. Non-inclusion virus of citrus red mite; negatively stained virion (arrow); ×450,000. (Photo by K. SMITH, courtesy H. TASHIRO)

Distribution of Viruses Among Arthropods

Over 250 arthropod species have at least one virus disease, and about 300 "virus species" are described (Table 3). Individual insect species may have from 1 to 4 different viruses. Approximately 85 percent of arthropod viroses were isolated from caterpillars of moths and butterflies. This group (order-*Lepidoptera*) contains many economically important pests. Ants, bees, and wasps (order *Hymenoptera*) account for ten percent of the arthropod viruses. The remaining five percent is about equally divided between two

Table 3. *Kind and number of arthropod viruses*

Arthropod host		Type of Arthropod Virus					
Common name	Order	Nuclear polyhedrosis	Cytoplasmic polyhedrosis	Granulosis	Polymorphic inclusion	Non-inclusion	Total
Moths, Butterflies	Lepidoptera	105	85	37	3	3	233
Ants, Bees, Wasps	Hymenoptera	12	12	0	0	3	27
Flies, Mosquitoes	Diptera	4	2	0	0	5	11
Beetles	Coleoptera	0	0	0	1	2	3
Grasshoppers	Orthoptera	1	0	0	0	1	2
Lace wings	Neuroptera	0	2	0	0	0	2
Spider mites	Acarina	0	0	0	0	3	3
	Total	122	101	37	4	17	281

Fig. 15. Non-inclusion virus of crane fly, *Tipula spp.*; double shadowed at 60° α; icosahedral virion (arrow); ×160,000. (Courtesy of C. RIVERS and D. GLITZ)

Table 4. *Viruses isolated from selected arthropod pests*

Target pest	Plants, animals or goods attacked	Virus type
Armyworm complex	Food and fiber crops	Nuclear polyhedrosis, Cytoplasmic polyhedrosis, Granulosis, non-inclusion
Bollworm-budworm complex	Food and fiber crops	Nuclear polyhedrosis, Cytoplasmic polyhedrosis Granulosis
Cabbageworm	Food crops	Nuclear polyhedrosis, Cytoplasmic polyhedrosis, Granulosis, Polymorphic inclusion
Citrus red mite	Fruit trees	Non-inclusion
Clothes moth	Household products and goods	Nuclear polyhedrosis, Cytoplasmic polyhedrosis
Mosquito complex	Man and animals	Nuclear polyhedrosis, Non-inclusion
Sawfly complex	Forest and ornamental trees	Nuclear polyhedrosis, Granulosis
Tent caterpillar complex	Forest, ornamental and shade trees	Nuclear polyhedrosis

groups — the flies-mosquitoes (order *Diptera*) and the bettles-grasshoppers-lace wings- and spider mites (orders-*Coleoptera, Orthoptera, Neuroptera,* and *Acarina*). The present frequency distribution of arthropod viruses is a reflection of entomologists concentrating on economically important arthropods and the relative ease of detecting inclusion type viroses. As other invertebrates are studied, more viruses are being reported from these groups. Some as the result of natural infections, others by experimental cross-transmission using viruses from other insects. The *Tipula* irridescent non-inclusion virus of the crane fly (Fig. 15) (*Tipula paludosa*), for example, was experimentally transmitted (mainly by injection) to 21 insects (SMITH et al., 1961).

Selected viruses of major arthropod pests are presented in Table 4. Intelligent selection of a commercial viral insecticide requires knowledge of the virus-pest association.

Nomenclature and Classification of Arthropod Viruses

Short phrases are often used to describe arthropod viruses in a manner similar to that commonly used to describe plant and animals prior to the development of the Linnean system of nomenclature. HOLMES (1948) first named and classified animal and plant viruses, using the binomial concept. HOLMES placed insect viruses into two genera, the *Borrelina* viruses (polyhedral wilt and other diseases), and the *Morator* virus (sac brood of honey bees). STEINHAUS (1949) later proposed four major groups which he called *natural groups* and raised these groups to the rank of genus. These were: *Borrelina* (polyhedral inclusions), *Paillotella* (polymorphic inclusions), *Bergoldia* (granulosis), and *Morator* (non-inclusion viruses). STEINHAUS wisely also suggested that the actual naming of species be delayed until definte information on morphology of the virus particle became available. Additional modifications of the basic groups were later suggested by STEINHAUS (1953), BERGOLD (1953a), ZHDANOV (1953), WEISER (1958), and VAGO (1958). An insect virus study group was eventually formed under the auspices of the "Virus Subcommittee on Bacteriological Nomenclature". The virus subcommittee could not agree on a uniform system but did propose the suffix *virus* be added to each generic name. This was accepted by the insect virus study group in combination with a virus species and author designation (BERGOLD et al., 1960). Nine genera now exist for all the known arthropod viruses.

The use of a Latinized binomial is commendable. It fits in well with nomenclature generally established for plant and other animal viruses. However, a more systematic approach to arthropod virus nomenclature and classification based on the virion must be established before the situation becomes unwieldy. Arthropod viruses must not be left in a nomenclature-taxonomic vacuum. Only three criteria are consistently used to name and/or classify arthropod viruses: (1) Inclusion body presence or absence, (2) inclusion body shape, and (3) site of viral replication. The most important basis of any system — the organism — in this instance the virion, is largely neglected when naming or classifying arthropod viruses. Criteria such as virion morphology, virion chemistry and nature of the nucleic acid, virion genetic, virion symmetry, and virion antigenicity suggested by LWOFF and TOURNIER (1966) for vertebrate viruses could and should be used to name and classify arthropod viruses.

Evaluation of Arthropod Viruses as Control Agents

The questions, "What makes a good viral insecticide?" cannot be categorically answered. Criteria used to evaluate chemical insecticides are starting points but it should be remembered that these were established for "non-living" chemicals and not "living" biologicals. An overall evaluation of arthropod viruses as viral insecticides can be based upon three major points: (1) Known or potential effectiveness against economically important pests, (2) safety to humans, other vertebrates, beneficial invertebrates and plants, and (3) feasibility of commercial production.

Known or Potential Effectiveness

Various questions must be asked and answered under known or potential effectiveness (Table 5). Is the virus naturally associated with the target pest? Does the virus kill as quickly as granuloses and nuclear polyhedroses, or is it a debilitive type virus as exemplified by the cytoplasmic polyhedroses? What is the anticipated effective field dose? How and when should the virus be applied to insure maximum effectiveness? Is the treated population resistant to virus? Will it develop resistance after continued re-application? Are

Table 5. *Known or potential effectiveness of several types of arthropod viruses*

riteria	Pine processionary moth Cytoplasmic polyhedrosis[a]	Bollworm Nuclear polyhedrosis[b]	Codling moth Granulosis[c]	Tipula irridescent Non-inclusion[d]
ype of Activity	debilitive	lethal	lethal	debilitive
ethal dose LD^{50}/gram insect				
Inclusion bodies	28×10^5	200—8,000	13×10^6	—
Virions $\times 10^3$	14×10^5	2—8	$13—26 \times 10^6$	8—40
ethal time — Days LT^{50}	7	3—6	3	14—28
ield dose — Billions				
nclusion bodies/acre	324	60—600	500	undetermined
esistance development	not reported	not reported	not reported	not reported
lost specificity	order	generic	family	order
nvironmental persistence	good	moderate to good	not determined	poor
gricultural compatibility	good	good	good	not evaluated

[a] MARTOURET et al. (1957); GRISON (1960, 1967).
[b] IGNOFFO (1965a, b).
[c] FALCON (1967); TANADA (1964, 1967).
[d] RIVERS and GLITZ (1967); SMITH and HILL (1962).

viruses specific? What will deactivate field-applied virus and how can this be overcome to increase viral effectiveness? And, last, is the virus compatible with presently used agricultural chemicals and agricultural practices and is it stable under manufacturing and storage conditions? Of course, it is impossible to answer all these questions, nor am I implying that all the information is available to do so. It is hoped that a tabulated comparison of major types of arthropod viruses (Table 5) and the discussion which follows will give the reader some idea of the known or potential effectiveness of arthropod viruses.

Field Use of Arthropod Viruses. The literature cites over 30 examples from different parts of the world in which arthropod viruses were used to control economically important pests. These examples are cited in the reviews of SMITH and WILLIAMS (1958), TANADA (1959), FRANZ (1961), CAMERON (1963), VAGO (1964), HEIMPLE (1965), and IGNOFFO (1966a, 1967). All major viral groups, except the polymorphic inclusions are represented (Table 6). The nuclear polyhedroses and granuloses, probably because of their virulence, were used more than any other virus.

Arthropod viruses are naturally present in many agricultural and forest environments. Cabbage loopers, *Trichoplusia ni*, cannot be fed raw cabbage

Table 6. *Selected examples of viruses applied to control arthropod pests*

Virus type and target pest		Literature citation
Nuclear polyhedroses		
Alfalfa caterpillar	*Colias eurytheme*	STEINHAUS and THOMPSON, 1949
Cabbage looper	*Trichoplusia ni*	HALL, 1957
Cotton bollworm	*Heliothis zea*	IGNOFFO et al., 1965
Cotton leafworm	*Prodenia litura*	ABUL NASR, 1959
European pine sawfly	*Neodiprion sertifer*	BRENY, 1951
European spruce sawfly	*Diprion hercyniae*	BALCH and BIRD, 1944
Gamma noctuid	*Plusia gamma*	VAGO and CAYROL, 1955
Great Basin tent caterpillar	*Malacosoma fragile*	CLARK and THOMPSON, 1954
Gypsy moth	*Porthetria dispar*	ROLLINSON et al., 1966
Spruce budworm	*Choristoneura fumiferana*	STAIRS and BIRD, 1964
Swaine jack-pine sawfly	*Neodiprion swainei*	SMIRNOFF, 1961
Wattle bagworm	*Kotochalia junodi*	OSSOWSKI, 1957
Cytoplasmic polyhedroses		
Pine processionary caterpillar	*Thaumetopoea pityocampa*	BILIOTTI et al., 1959 GRISON, et al., 1959
Granuloses		
Cabbageworm	*Pieris brassicae*	BILIOTTI et al., 1956
Fall webworm	*Hyphantria cunea*	FRANZ, 1961
Imported cabbageworm	*Pieris rapae*	TANADA, 1953; KELSEY, 1958
Red-banded leaf roller	*Argyrotaenia velutinana*	GLASS, 1958
Spruce budworm	*Choristoneura fumiferana*	STAIRS and BIRD, 1964
Non-inclusion viroses		
Citrus red mite	*Panonychus citri*	GILMORE, 1965

or lettuce bought at the food mart. Enough nuclear polyhedrosis virus is present in a lettuce salad to kill over a million young cabbage loopers. If the virus is naturally present and at such high levels, why is it necessary to apply virus in order to control a pest like the cabbage looper? The answer is related to timing and the level of acceptable damage. Natural epizootics occur after the pest has done considerable damage. Therefore, to prevent or minimize

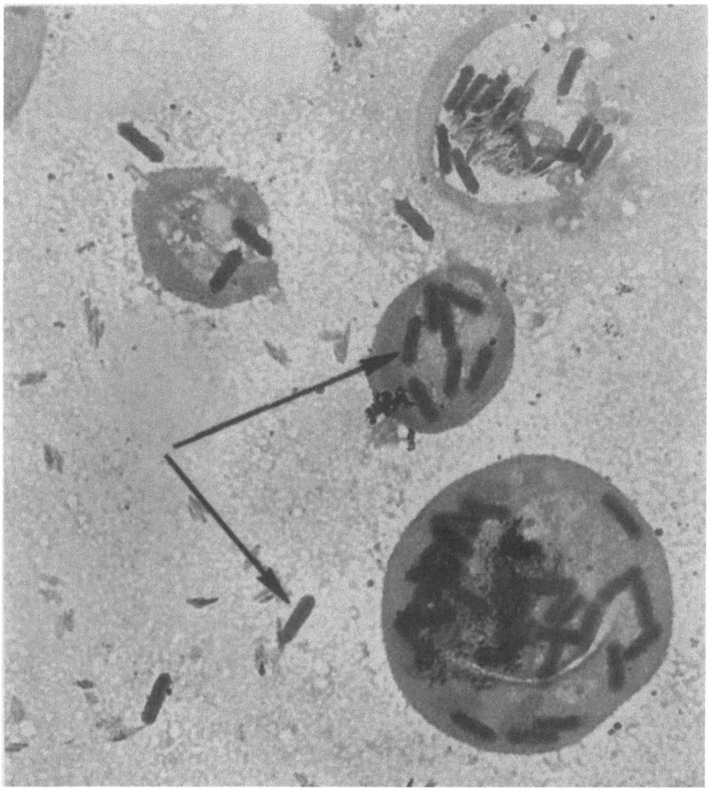

Fig. 16. Nuclear polyhedrosis virus of corn earworm; partially dissolved inclusion bodies with released virions (arrow); ×25,000. (Photo by B. GREGROY)

damage on crops, the virus must be applied early. Young cabbage looper caterpillars after hatching feed on virus sprayed plants, swallow the inclusion bodies which dissolve in the midgut releasing infectious viral units (Fig. 16). These units pass through the epithelial cells, which line the midgut, into the hemocoele, and eventually penetrate the cytoplasm and nuclei of susceptible cells. Replication and formation of virions and inclusion bodies are in the nuclei of susceptible cells. Most susceptible cells are invaded within 2 to 7 days after initial infection. Cell lysis and disintegration of larval tissue occurs shortly thereafter.

Viruses can be used as either density-independent control agents or as density-dependent control agents. In the density-independent approach, application is by conventional sprayer, bait-applicators or dusters (Fig. 17).

The insecticidal approach is the most commonly used technique. It fits current agricultural practices but is both costly and wasteful for biological insecticides. In the density-dependent approach, the virus is introduced into a population and allowed to spread much like beneficial parasites are introduced and colonized. Introduction and colonization was used by GILMORE (1965) and later GILMORE and TASHIRO (1966) to evaluate the potential effectiveness of the non-inclusion virus of the citrus red mite, *Panonychus citri* (Fig. 18). It is

Fig. 17. Aircraft application of cotton bollworm, *Heliothis zea*, nuclear polyhedrosis virus. (Photo by J. L. KNAPP)

less costly than the insecticidal approach but inherently slow acting and, consequently, will be used only where higher damage levels can be tolerated.

Careful consideration must also be given to when and how virals are applied in the field (THOMPSON and STEINHAUS, 1950; MCEWEN and HERVEY, 1958). CHANT (1964), BRAZZEL (1965), and ALLEN et al. (1966) suggest an integrated control approach. This takes full advantage of the natural presence of parasites and predators which are normally preserved in virus treated fields (FALCON et al., 1965; LEIGH et al., 1966).

Consistent, uniform plant coverage with virus is also of paramount importance. Mode of application, application initiation, re-application interval, and number of applications established for chemical insecticides will have to be re-evaluated and optimized for biological insecticides. Theoretically, 10 gallons of water/acre applied to a perfectly flat surface deposits over two thousand 100 micron drops/centimeter^2, with drops spaced 0.2 mm apart (CARLETON et al., 1966). Increasing the drop size to a diameter of 1,000 microns, decreases the deposit to 2 drops/centimeter^2 but more importantly, the drops are now spaced 7 mm apart. How would this relate to a field dose of nuclear polyhedrosis virus for control of cotton bollworms? Ten gallons of water contain about 5×10^{11} inclusion bodies. At 100 micron-diameter, each drop contains about six inclusion bodies. The LD_{50} for young cotton bollworms is less than 3 PIB/mm^2 (IGNOFFO, 1966b).

Control of sawflies, *Diprion* and *Neodiprion* spp. (BIRD, 1953; BIRD and BURK, 1961), alfalfa caterpillar, *Colias eurytheme* (THOMPSON and STEINHAUS,

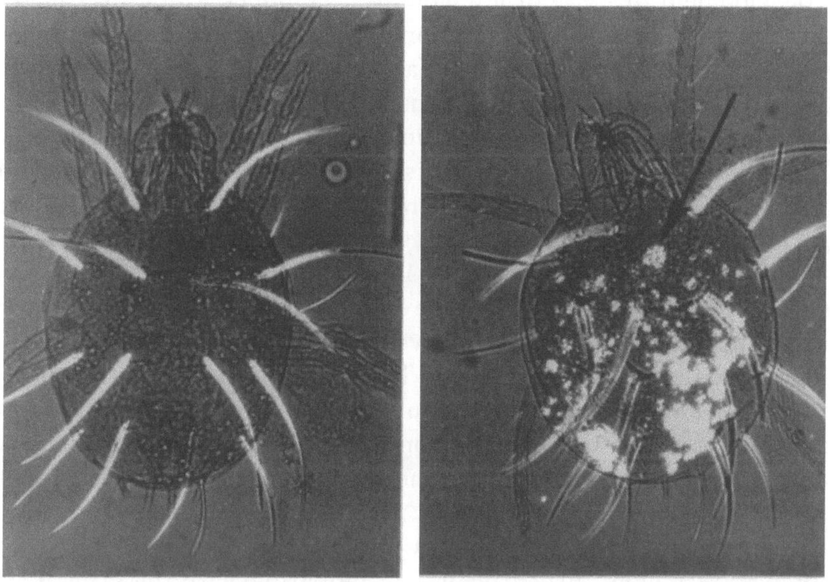

Fig. 18. Non-inclusion virus of citrus red mite; healthy (left) and diseased (right) mites. Birefringent crystals (arrow) are disease manifestation and not virus *per se*. (Courtesy H. TASHIRO)

1950), cabbage looper (ELMORE, 1961; GENUNG, 1960), cotton bollworms and tobacco budworms, *Heliothis zea* and *Heliothis virescens* (IGNOFFO et al., 1965) with sprays of the nuclear polyhedrosis virus; cabbageworms, *Pieris brassicae*, *Pieris rapae* (TANADA, 1956a, b; KELSEY, 1957; BILOTTI, 1956) with granulosis sprays; and the pine processionary moth, *Thaumetopoea pityocampa*, with a cytoplasmic polyhedrosis dust (BILIOTTI et al., 1959a, b, c) are some examples where good to excellent control was obtained. Baits also are used to increase effectiveness of field applied virus, much as they are used to increase effectiveness of chemicals (SHAW, 1955) and microbial insecticides (MCLAUGHLIN, 1966). Several laboratory studies have shown that plant extracts mixed with viruses will produce a higher incidence of virus than virus without the extract (MONTOYA, et al., 1966; ALLEN and PATE, 1966).

The feeding and behavioral habits of the target pest on a particular host-crop is another important factor to consider when evaluating the effectiveness of field applied viruses. Bollworms feeding "exposed" on lettuce present a different control problem than bollworms feeding "hidden" on cotton, corn, or tomatoes. Cabbage loopers feed continuously on exposed cotton or cabbage foliage where they are more likely to ingest virus. Under some conditions, the type of host plant (DAVID & GARDINER, 1965a, b) or the population density may (JAQUES, 1962a; ARUGA, 1963) or may not (STEINHAUS, 1958; DAVID & GARDINER, 1965a) stress an increase in the incidence of granuloses or nuclear polyhedroses. Incidence of a nuclear-polyhedrosis was about four times greater under conditions of "physiological stress" than under "normal" conditions (JAQUES, 1961).

The development and use of low volume techniques which maximize crop coverage; more effective sampling techniques to initially detect presence of target pests; and the integrated control approach should certainly increase the overall effectiveness of field applied viruses.

Agricultural Compatibility. As previously stated, virus must be ingested by the target pest if economic control is to be achieved. Adjuvants are combined with chemical insecticides to increase field effectiveness by increasing coverage, adherence, and persistence. Conceivably viral effectiveness may increase if the virus is combined with insecticides or insecticidal adjuvants. Viruses should therefore be compatible with agricultural chemicals, such as insecticides and fungicides, and fit conventional agricultural practices (application techniques, crop programming). Previous studies indicated that most chemical insecticides or insecticidal adjuvants are compatible with arthropod viruses. However, tests were usually conducted under field conditions where slight effects are undetectable unless the materials are grossly incompatible.

SMIRNOFF (1961) laboratory tested 28 insecticides or adjuvants for compatibility with the nuclear polyhedrosis virus of the Swaine jackpine sawfly, *Neodiprion swainei*. Only turpentine was incompatible with the virus. IGNOFFO and MONTOYA (1966) tested six insecticides and six adjuvants for compatibility with the nuclear polyhedrosis virus of the cotton bollworm. Only methyl-parathion deactivated the virus. Many oils, insecticides, and surfactants are compatible with the nuclear polyhedrosis virus of *Heliothis* (TANADA and REINER, 1962; WOLFENBARGER, 1964, 1965; IGNOFFO et al., 1965). Similar results of compatibility were obtained for the nuclear polyhedrosis virus of the alfalfa caterpillar (CLARK and REINER, 1956), and for the cabbage looper (GENUNG, 1960; ELMORE, 1961; GETZIN, 1962; HOFMASTER and DITMAN, 1961; IGNOFFO, 1964; JAQUES, 1961, 1962; MCEWEN and HERVEY, 1958, 1959; WOLFENBARGER, 1965). Oil + endrin gave better results than a threefold increase in the virus concentration (WOLFENBARGER, 1965).

Surfactants did not adversely affect the granulosis virus of the codling moth, *Carpocapsa pomonella* (TANADA, 1964) or the imported cabbageworm, *Pieris rapae*, (TANADA, 1956b; MCEWEN and HERVEY, 1959). The granulosis virus of the red-banded leaf roller, *Argyrotaenia velutinana*, was compatible with five fungicides (captan, dichlone, ferbam, phenyl mercuric acetate, and sulfur) but not with glyodin (GLASS, 1958).

Inclusion bodies and virions are generally deactivated by pH values below 5 and above 8 (BERGOLD, 1958, 1963; SMITH, 1963; HUGER, 1963; IGNOFFO and

GARCIA, 1966). To minimize these effects, field applied viruses should be buffered to maintain near-neutral pH. In some instances this may prove difficult, because of extreme variations in water pH from different geographical areas.

Very little effort has been devoted to preparing insect viruses for field use. This is quite surprising since the ultimate success or failure of field tests can be related to how the virus was handled during collection, purification, and storage. Dried inclusion viruses are very stable and will survive normal storage temperatures ranging from several months to many years (BERGOLD, 1958; NIELSON and ELGEE, 1960). The nuclear polyhedrosis virus of the silkworm was still infectious after 20 years storage (mostly refrigerated) in a flame-sealed glass tube (STEINHAUS, 1960a, b).

Environmental Persistence. Discussions on persistence of virus include the period during which control is desired within a particular crop season, as well as long-term survival which extends from season to season. Environmental factors, i.e., extreme temperature, rainfall, light, free-water, pH, and plant interactions may deactivate field applied viruses. These factors not only affect viral viability and infectivity, but also the pest's feeding rate on treated foilage. Environmental "stress" factors (heat, cold, ultraviolet light) can also influence the incidence of viral diseases, e.g., granulosis of the variegated cutworm, *Peridroma margaritosa*, (STEINHAUS and DINEEN, 1960). Inclusion viruses will probably persist longer than non-inclusion viruses.

Most examples of long-term persistence are with viruses of forest insects. Virus in these examples survive in living adults (BIRD, 1961; BIRD and BURK, 1961) or in diseased cadavers (CLARK, 1955; OSSOWSKI, 1960a, b). Viruses may also persist as a "sub-acute" or "latent" infection in overwintering stages of arthropods. It is more probable that virus applied to forests persists longer than virus applied to field crops.

Field applied viruses persist in soil for several years (STEINHAUS and THOMPSON, 1949). Alfalfa caterpillars, exposed to soils from alfalfa fields, developed typical symptoms of nuclear polyhedrosis. Soil inoculated with nuclear polyhedrosis virus of the cabbage looper retained its infectivity over a period of two years and was detected on cabbage foilage after heavy rains (JAQUES, 1964).

Ultraviolet light will deactivate viruses under laboratory conditions (WATANABE, 1951; AIZAWA, 1955) but the role of UV and/or sunlight, in deactivating field applied virals on leaf surfaces, needs further clarification. Sunlight-UV, although not directly responsible, may indirectly activate or catalyze reactions which result in viral instability and loss of infectivity (HIRT, et al., 1960; TURNER and KAPLAN, 1965).

Laboratory tests indicate that normal field temperatures (10—30°C) should not adversely inhibit viral activity or viral development in diseased caterpillars. Continual exposure at higher field temperatures (35—45°C) may affect virus stability and inhibit viral multiplication (BIRD, 1955; THOMPSON, 1959; IGNOFFO, 1966c). The thermal inactivation of inclusions and their enclosed virions under laboratory testing is not different from the thermal deactivation of most proteins (BERGOLD, 1953b; WATANABE, 1951; AIZAWA, 1953a, b). High or low field temperatures do affect ingestion of virus treated plants and over longer periods may affect the incidence of viral infections.

Field applied viruses are probably not adversely affected by high humidities. High humidity was related to spread and development of natural viral epizootics in populations of the gypsy moth, *Porthetria dispar*, (WALLIS, 1957) oakworms, *Phryganida californica*, (SCHMIDT and PHILLIPS, 1958) and the gamma noctuid, *Plusia gamma*, (VAGO and CAYROL, 1955) but not to natural viral epizootics of sawflies, *Neodiprion* and *Diprion* spp. (BIRD and ELGEE, 1957; KRIEG, 1955).

Under natural conditions of weathering, the granulosis virus of the cabbageworm persisted for four months on treated cabbage plants and was not lost when exposed for five hours to a simulated rainfall of 100 inches (DAVID and GARDINER, 1966a). The virus even persisted when cabbage leaves were "scrubbed" with a detergent or "treated" with caustic potash. BURGERON and GRISON (1965) obtained similar results with the cytoplasmic polyhedrosis of the pine processionary moth.

Resistance of Arthropods to Virus. Insects develop resistance to most of the more widely used chemical insecticides, therefore, entomologists are often asked whether insects will develop resistance to field applied viruses. We know from other biological systems that strains of animal, plant, and bacterial viruses exist and that host strains are differentially susceptible to a particular virus strain. Insects are probably no different in this respect than other living organisms. Neonate tobacco budworms were 30 to 50 percent more susceptible than cotton bollworms (IGNOFFO, 1966b) and one species of forest tent caterpillars, *Malacosoma disstria*, could be infected with the nuclear polyhedrosis isolated from three other species, *M. americanum*, *M. pluviaale*, and *M. alipicola* (STAIRS, 1965).

Resistance was never experimentally induced in a laboratory population of insects, although virus-resistant strains of both natural and laboratory populations of insects were reported (OSSOWSKI, 1957, 1958, 1960a, b; RIVERS, 1959; DAVID and GARDINER, 1960, 1966b; SMIRNOFF, 1961; ARUGA and WATANABE, 1964; WATANABE, 1965). Conversely, there are still reports of epizootics of nuclear polyhedrosis virus in natural populations of alfalfa caterpillars or cabbage loopers (STEINHAUS, 1949; IGNOFFO, 1964). A polyhedrosis virus held natural populations of the European spruce sawfly, *Diprion hercyniae*, under control for over 20 years with no detectable development of resistance (BIRD and BURK, 1961). CAMERON (1963) and MARTIGNONI and SCHMIDT (1961) suggest that fluctuations of natural populations of insects may be related to continuous selection of strains of both the insect and the virus.

Larval maturation is cited to explain the development of resistance in some granuloses and polyhedroses (CLARK and THOMPSON, 1954; TANADA, 1956a, b; HUGER, 1963; STAIRS, 1965). STAIRS (1965) found that the resistance of forest tent caterpillars increased as the caterpillars matured. The estimated

LD^{50} for a 1st instar larvae was 50 inclusion bodies, while that of 4th instar larvae was 2 million inclusion bodies. Others suggest that older larvae are more susceptible or that the normal increase in body weight "dilutes" a constant viral dose (KRIEG, 1955; SMITH et al., 1956; IGNOFFO, 1966d). Sawfly prepupae are the only stage not susceptible to viral infections by polyhedrosis (BIRD, 1953; BIRD and WHALEN, 1953). Presumably, the embryonic cells of the prepupal midgut is more resistant than cells of the other stages. It was generally assumed that pupae and adults are resistant to viral infections. However, VAGO (1957), MARTIGNONI & SCHMIDT (1961), and more recently, VAIL (1967) demonstrated that both pupae and adults are susceptible to nuclear polyhedroses viruses. NIELSON (1965) obtained similar results with a cytoplasmic polyhedrosis virus.

It may be possible to induce a level of resistance by temperature or chemicals. Temperatures above 30°C significantly increase the resistance of the wax moth larvae, *Galleria mellonella,* to intrahemacoelic injection of the non-inclusion virus of the crane fly (TANADA and TANABE, 1965). Temperature-induced-resistance is also reported by BIRD (1955) and THOMPSON (1959). Recently, BEN-SHAKED and HARPAZ (1966) were able to induce resistance in the cotton leafworm, *Prodenia litura,* (nuclear polyhedrosis virus) after injection with ether extracts of healthy insects.

Specificity of Arthropod Viruses. Insect virologists originally believed that virus-host specificity was a general phenomenon. This view, first challenged by SMITH and XEROS (1953), STEINHAUS (1953), GERSHENEON (1955), and AIZAWA (1962) is not universally accepted today. As more information is obtained, levels of viral specificity are moved to generic, familial, or ordinal levels. Most arthropod viruses are probably "specific" on generic or supra-generic levels.

Six species of *Heliothis* (*H. obtectus, H. virescens, H. zea, H. phloxiphaga H. armigera,* and *H. paradoxus*) are all susceptible to one isolate of the *Heliothis* nuclear polyhedrosis virus and virus isolated from *Heliothis peltigera* (HARPAZ & ZLOTKIN, 1966) will infect three of the above species (IGNOFFO, 1965a). More recent data show that virions of the *Heliothis* nuclear polyhedrosis do not, when injected, infect the cabbage looper, wax moth, lucerne moth (*Nomophila noctuella*) or housefly (*Musca domestica*). NIELSON (1964) experimentally transmitted the cytoplasmic polyhedrosis virus of the painted lady butterfly, *Vanesa cardui*, to 11 *Lepidoptera* and 20 *Hymenoptera*. IGNOFFO and ADAMS (1966) successfully transmitted a cytoplasmic polyhedrosis of the pink bollworm (*Pectinophora gossypiella*) to the cabbage looper, cotton bollworm, tobacco budworm, but not to the wax moth, southern armyworm (*Prodenia eridania*), European corn borer (*Ostrinia nubilalis*), lobster roach (*Nauphoeta cinerea*), American cockroach (*Periplaneta americana*), or the housefly. A recently described "polymorphic-type" inclusion virus (WEISER and VAGO, 1966) could not be transmitted to the fall webworm (*Hyphantria cunea*), gypsy moth, or to *Euproctis chrysorrhoca.* The non-inclusion virus of the crane fly was experimentally transmitted by injection to 7 species of *Diptera,*

11 species of *Lepidoptera*, and 3 species of *Coleoptera* (SMITH, et al., 1961). Until recently, only the granuloses have maintained species-specificity; however, even this is now questionable. Dr. L. FALCON (1967) of the University of California, Berkeley, recently communicated that he was able to transmit the codling moth granuloses to the oriental fruit moth, *Grapholitha molesta*.

Natural Transmission of Arthropod Viruses. The most common method of virus spread among insects is by ingestion of contaminated materials, such as food, egg shell, feces, or other diseased larvae. Pathologists generally agree that contact itself is not sufficient to transmit the disease from diseased to healthy insects in nature, although contact-transmission is possible in the laboratory (JAQUES, 1962b). Stings of parasitic wasps (*Apanteles, Bracon*) may also mechanically transmit virus.

Virus survival and transmission from one stage to another across generations (transtadial) is probably due to virus carried on or inside the egg. Most workers studying transtadial transmission fail to experimentally differentiate between transovum, i.e., general transmission via eggs and transovarial, i.e., transmission within the ovary. Transovarial transmission would increase the probability of persistence in natural populations of arthropods. External egg contamination can partially explain the "spontaneous" development of field epizootics of viruses. Adult cabbage loopers, artifically contaminated with nuclear polyhedrosis virus, mechanically transmit the virus to eggs and/or larvae (ELMORE and HOWLAND, 1964); however, dispersion by foliage application was more effective than adult dispersion. In an earlier study, MARTIGNONI and MILSTEAD (1962) demonstrated that adult alfalfa caterpillars could transmit virus to eggs and/or young larvae if the ovipositor is artifically contaminated.

True transovarial transmission is supported by evidence of SAGER (1960) and VAIL (1967). SAGER observed virus-like bodies in eggs obtained from "diseased" adults, and concluded that transtadial transmission of nuclear polyhedrosis or granulosis occurs in seven species of *Lepidoptera* which he studied (*Peridroma margaritosa, Colias eurytheme, Bombyx mori, Hemerocampa pseudotsugata, Orgyia antigua, Lambdina fiscellaria, Pseudaletia unipuncta*). However, he felt that in some insects, transtadial-transmission was the exception rather than the rule. Recently, VAIL (1967) detected virus inclusions associated with ovaries of cabbage looper pupae previously injected with virions of the nuclear polyhedrosis virus.

Generation to generation transmission via the egg was demonstrated for nuclear polyhedroses of *Prodenia litura* by HARPAZ and BEN-SHAKED (1964), *Neodiprion swainei* by SMIRNOFF (1962), and for the cytoplasmic virus of the silkworm (ARUGA and NAGASHUMA, 1962; ARUGA, 1963; HUKUHARA, 1962).

Pathologists also speculate on the spontaneous occurrence of diseases in healthy insects (BERGOLD, 1958; STEINHAUS, 1956, 1958, 1960a). It is suggested that these insects carry "latent" infections or develop infections from "occult" viruses. An occult virus has been defined as an undetectable viral stage. Disease symptoms of virus only appear after the host is environmentally "stressed." The occult virus of arthropods is probably analogous to pro-phage or pro-virus in plants, bacteria, and higher animals. DARLINGTON (1958) proposed a scheme which may be applicable to occult arthropod viruses. Particles called plasmagenes (occult viruses) are capable of transforming under appropriate conditions into infectious virus. The appropriate conditions in the case of arthropod virus may be outside environmental factors (STEINHAUS, 1958). Carbon dioxide stress of fruit flies is often cited as a specific example of an occult virus (L'HERITIER, 1948). Carbon dioxide sensitivity can be transmitted by injection to resistant fruit flies. The causative agent was previously classified as a virus on the basis of its size, course of multiplication, and ability to undergo mutation. Its existence as a non-inclusion rod, was recently confirmed by electron microscopy (BERKALOFF, et al., 1965). BERGOLD (1958) suggested that latent infections are present in both natural and laboratory populations of arthropods and that field application of virus produce control through *super-activation* of these infections. No evidence of latent infections was observed by BIRD (1961) in either laboratory or field population of sawflies. The transtadial and transovarial development of vertebrate diseases in tick and mite vectors was recently reviewed in detail by BURGDORFER and VARMA (1967).

Safety to Animals and Plants

Since viruses will be used to control field pests and are able to persist in both soil and plant environments, they must be registered with the United States Department of Agriculture (USDA) before they can be sold in interstate commerce. The USDA, Agricultural Research Service (ARS) has the responsibility of administering provisions of the Federal Insecticide, Fungicide, and Rodenticide Act passed in 1947. Viral insecticides must be effective against the target pest and safe to humans, livestock, wildlife, and crops before they can be registered. In addition, the law requires manufacturers to provide research data on whether residues will remain on food or livestick products at harvest.

Safe limits called "tolerances" which prescribe the amount of residue levels permitted on crops are established by the Food and Drug Administration (FDA) of the U.S. Department of Health, Education, and Welfare, under the provisions of the Federal Food, Drug and Cosmetic Act of 1938, and the Miller Pesticide Chemical Amendment of 1954. Another branch of government, the U.S. Department of Interior, also reviews the registration-application for its possible role in human health and wildlife conservation.

Complete exemption from all tolerance levels is the objective sought for viral insecticides. To obtain the necessary information, data is accumulated on the presence or absence of activity as it relates to possible toxicity, pathogenicity, allergenicity, carcinogenicity, and teratogenicity. Vertebrate test animals are exposed to acute and chronic doses by feeding, inhalation, and injection into the skin, muscles, peritoneum, and the circulatory and nervous system. Possible persistence of the virals in tissue, blood, or feces is also considered. SMIRNOFF & MACLEOD (1964) could not detect virus (microscopically or biologically) in feces after feeding white mice the nuclear polyhedrosis virus of the wax moth, Swaine jack-pine sawfly, and linden looper (*Erannis tillaria*). None of the mice show any adverse effects.

Insect virologists believe viruses are safe and probably less objectionable than most chemical insecticides (Table 7). Arthropod viruses were commercially and experimentally applied on vast acreages of food, forage, fiber crops, forest, fruit, and ornamental trees with not *one* reported case of toxicity or pathogenicity to man, warm-blooded animals or plants. Honey bees are not affected (POLTEV, 1963; CANTWELL et al., 1966) by granulosis or nuclear polyhedrosis viruses nor are beneficial parasites and predators (FALCON et al., 1965). In several instances, parasites and predators were observed developing normally on hosts infected with virus (TANADA, 1956b; BILIOTTI, 1956).

Table 7. *Safety of several types of arthropod viruses to plants, beneficial arthropods, vertebrates and humans*

Criteria	Pine processionary moth Cytoplasmic polyhedrosis[a]	Bollworm Nuclear polyhedrosis[b]	Codling moth Granulosis[c]	Tipula irridescent Non-inclusion[d]
Phytotoxicity-pathogenicity	negative	negative	negative	negative
Beneficial arthropod effect	negative	negative	undetermined	negative
Vertebrates and humans				
Toxicity	not reported	negative	undetermined	negative
Pathogenicity	not reported	negative	undetermined	negative
Allergenicity	not reported	negative	undetermined	undetermined
Carcinogenicity	undetermined	negative	undetermined	undetermined
Teratogenicity	undetermined	negative	undetermined	undetermined

[a] MARTOURET et al. (1957); GRISON (1960, 1967).
[b] IGNOFFO (1965a, b).
[c] FALCON (1967); TANADA (1964, 1967).
[d] RIVERS and GLITZ (1967); SMITH and HILL (1962).

The safety of viruses to warm-blooded vertebrates is demonstrated both by indirect and direct evidence. Natural viral epizootics in laboratory and field population historically verified that insect virus diseases are limited to the target pest. Evidence compiled by many scientists showed that six vertebrate species (3,000+ animals) were treated with 28 different insect viruses without any demonstrable case of viral toxicity, pathogenicity, or allergenicity. Within the last three years, specific evidence on the safety of two insect viruses to vertebrates was obtained. The cotton bollworm nuclear polyhedrosis virus was administered as either inclusion bodies, virions, or polyhedral-protein by inhalation, feeding, and intravenous, intradermal, intraperitoneal, and intracerebral injections to 230 guinea pigs and/or white mice. Only one death occurred in a guinea pig exposed to virus dusts. The cause of death was later diagnosed as an acute pneunomia (IGNOFFO and HEIMPEL, 1965). No adverse effects were noted by HEIMPEL (1966) when guinea pigs and white mice were tested with virions or inclusion bodies of the nuclear polyhedrosis virus of cabbage loopers. Inclusion bodies containing infectious *Heliothis* virus were also fed to ten men and women over a period of five days with no harmful results (HEIMPEL and BUCHANAN, 1967). Each person consumed close to six billion inclusion bodies or enough virus to treat 0.1 to 0.01 acres of cotton (IGNOFFO et al., 1965; CHAPMAN et al., in press).

Production and Industrial Feasibility

Viral insecticides to be bought and used by growers must be effective, economical, and offer a distinct advantage over other insecticides. In most instances they will have to directly compete with commonly used chemical insecticides which have built up a "following", as well as a "functioning technology". Their availability then depends, to a great extent, on whether government of industry will invest both time and money into the development of a safe, selective viral insecticide (Table 8). The potential market of selective insecticides is restricted because of specificity to one target pest, or at best, a group of related pests (Table 4). Industrially, it is much easier to economically justify a virus which attacks many target pests (broad market potential) than one that attacks one target pest (narrow market potential). In addition, because of their specificity, insect viruses are difficult to produce on an industrial scale (Table 9). They are parasites and, so far as is known, can only be

Table 8. *Estimated dollar-cost to develop new insecticides*[a]

Developmental phases	Insecticide	
	Chemical Sevin[b]	Microbial Thuricide 90TS[c]
Synthesis and selection	500,000	80,000
Laboratory screening	200,000	480,000
Field testing	300,000	250,000
Toxicology	350,000	250,000
Patent	25,000	25,000
Formulation	80,000	100,000
Product development	450,000	355,000
Process development	350,000	275,000
Marketing and field sample	250,000	200,000
	2,505,000	2,015,000

[a] Development period 7 and 10 years, respectively.

[b] WELLMAN, R. L.: Industry's role in developing pesticides. Farm Chemicals **30**, 124—126 (1966).

[c] FISHER, R. A.: Personal communication 1966.

Table 9. *Propagation of various types of arthropod viruses*

Criteria	Pine processionary moth Cytoplasmic polyhedrosis[a]	Bollworm Nuclear polyhedrosis[b]	Codling moth Granulosis[c]	Tipula irridescent Non-inclusion[d]
Production:				
Technology	whole animal	whole animal	whole animal	whole animal
Substrate	foilage	synthetic diet	synthetic diet	foilage
Estimated propagation:				
Inclusions $\times 10^9$/gram	5—10	10—50	100	—
Virions $\times 10^{11}$/gram	20—50	1—5	1—2	200—700
Market	limited	good	moderate	limited

[a] MARTOURET et al. (1957); GRISON (1960, 1967).
[b] IGNOFFO (1965a, b).
[c] FALCON (1967); TANADA (1964, 1967).
[d] RIVERS and GLITZ (1967); SMITH and HILL (1962).

mass produced in living worms (Table 10). Thus, the living worms, which I lovingly call "crawling fermentors", are the only source of virus for field tests. Virus for field application has been collected from living and dead caterpillars in natural epizootics or from fields selectively treated to "produce" more virus. Laboratory or insectary rearing of caterpillars on contaminated foliage or injected pupae were also used (VAGO and ATGER, 1961). Some rather ingenious techniques, i.e., the *carrousel* method of SMIRNOFF (1964) were developed in an effort to mass-produce virus for field experimentation.

Nuclear polyhedrosis virus of the alfalfa caterpillar (STEINHAUS and THOMPSON, 1949), cabbage looper (MCEWEN and HERVEY, 1958; HOFMASTER and DITMAN, 1961), Great Basin tent caterpillar (CLARK and THOMPSON, 1954), and several species of sawflies (BALCH and BIRD, 1944; BIRD, 1950; LEWIS, 1960; FRANZ, 1961), cytoplasmic polyhedrosis of the pine processionary caterpillar (MARTOURET and DUSAUSSOY, 1959), and the granulosis virus of cabbageworms (BILIOTTI, et al., 1956; TANADA, 1953) were produced using the techniques described above. Caterpillars grown on a semisynthetic diet were used to produce nuclear polyhedroses of the cabbage looper and cotton bollworm (IGNOFFO, 1964, 1965a, b, c, 1966a), the cabbage armyworm, *Barathra brassicae* (PONSEN, 1966) and the granulosis virus of the codling moth (FALCON, 1967). Five insect viruses (Table 11), all nuclear polyhedroses, have been supplied by industry to field entomologists in order to evaluate their possible use as selective insecticides.

Within the last four years, considerable effort has been placed on the development and use on one of these viruses, i.e., the nuclear polyhedrosis virus of *Heliothis* spp. (IGNOFFO, 1965d). The virus infects the cotton bollworm, alias the corn earworm, tomato fruitworm (FIGS. 19, 20), and its closely related cousin, the tobacco budworm. The *Heliothis* virus was used on an estimated

Fig. 19. Healthy corn earworm, *Heliothis zea*, larvae feeding on ear of corn (actual size)

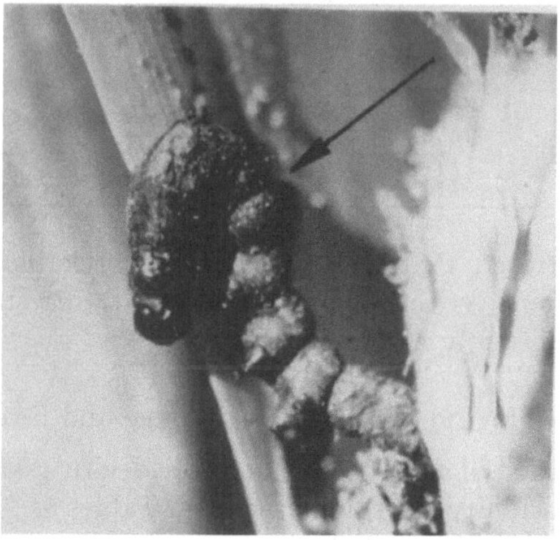

Fig. 20. Dead corn earworm, *Heliothis zea*, killed by its nuclear polyhedrosis virus; $\times 1\text{—}^1/_2$

50,000 acres of cotton, corn, sorghum, tomatoes, and tobacco. Most of this virus was manufactured and sold in southern Texas for use on cotton by HAYES-SAMMONS under the trade name VIREX. *Heliothis* virus was also made available for restricted use at state and federal experimental stations under the trade name VIRON/H^{tm} and BIOTROL-VHZ (Table 11). These extensive

Table 10. *Viruses grown in living insects or tissue-cell cultures*

Living insect culture	
Polyhedroses:	*Granuloses:*
Barathra brassicae	*Argyrotaenia velutinna*
Choristoneura fumiferana	*Carpocapsa pomonella*
Colias philodice eurytheme	*Choristoneura fumiferana*
Dendrolimus spectabilis	*Eucosma griseana*
Diprion hercyniae	*Laphygma exigua; Laphygma frugiperda*
Heliothis zea; Heliothis virescens; Heliothis phloxiphaga	*Pieris brassicae; Pieris rapae*
Kotochalia junodi	
Malacosoma fragile	*Trichoplusia ni*
Laphygma frugiperda; Laphygma exigua	
Neodiprion sertifer; Neodiprion swainei	*Non-inclusion viroses:*
Pectinophora gossypiella	
Peridroma margaritosa	*Panonychus citri*
Plusia gamma	*Tipula* irridescent
Porthetria dispar	*Sericesthis* irridescent
Prodenia litura; Prodenia eridania; Prodenia ornithogalli; Prodenia praefica	
Thaumetopoea pityocampa	
Trichoplusia ni	
Tissue-cell culture	
Non-inclusion viroses:	*Polyhedroses:*
Sericesthis irridescent	*Antheraea eucalypti*
Tipula irridescent	*Antheraea pernyi*
	Bombyx mori
	Hemerocampi leucostigma
	Galleria mellonella
	Peridroma magaritosa
	Samia walkeri

field trials demonstrated that the virus can be effective if used properly; however, its consistent, effective use, at least at this stage, requires more knowledge than is presently available.

Synopsis of Viral Effectiveness, Safety and Feasibility

A general evaluation of each type of arthropod virus, based upon effectiveness, safety and feasibility, is summarized in Tables 5, 7, and 9. Although obvious, it should be emphasized that no one virus is best for all target-pest host-plant systems. The nuclear polyhedroses and granuloses, because of their virulence, relative quick-activity, specificity, safety, stability, and industrial feasibility, are the most promising candiates for consideration as viral insecticides. Their use is similar to the use of chemical insecticides with the same ultimate objective. Less virulent debilitive-type cytoplasmic polyhedroses, which predispose the pest toward lethality, could be used to suppress pest populations over longer periods in a situation where growers and plants can tolerate higher levels of feeding damage. Experimental data is not available on the use of non-inclusion viroses as viral insecticides. Although virulent, non-inclusion viruses cannot be considered until their safety, effectiveness, and persistence under field conditions is demonstrated.

Although five polyhedroses were produced on a laboratory scale (Table 11), continual industrial production and economic use is yet to be conclusively established. In time, insect viruses may be successfully developed into industrial viral insecticides because they are effective, safe, and potentially feasible. Success is ultimately dependent upon overcoming the major obstacles of viral production and production cost.

Table 11. *Viruses made available by industry for field tests*

Target insect	Product name	Source
Bollworm-budworm complex	Biotrol VHZ	Nutrilite Products, Inc. — U.S.A.
	Viron/H tm	International Minerals & Chemical Corp., Chemical Division — U.S.A.
	Virex	Hayes-Sammons — U.S.A.
Sawfly complex	Polyvirocide	Indiana Farm Bureau Co.-Op. Assoc. — U.S.A.
Armyworms	*Prodenia* virus	International Minerals & Chemical Corp., Chemical Division — U.S.A.
	Biotrol-VPO	Nutrilite Products, Inc. — U.S.A.
Fall-beet army-worm complex	*Spodoptera* virus	International Minerals & Chemical Corp., Chemical Division — U.S.A.
	Biotrol VSE	Nutrilite Products, Inc. — U.S.A.
Cabbage looper	*Trichoplusia* virus	Biological Control Supplies — U.S.A.
	Trichoplusia virus	International Minerals & Chemical Corp., Chemical Division — U.S.A.
	Biotrol VTN	Nutrilite Products, Inc. — U.S.A.

Conclusion

Insect viruses, in general, and the nuclear polyhedroses, in particular, show considerable promise as living insecticides. They are effective, can be available and competitive as insecticides, are not poisonous nor do they cause diseases in man, other animals, beneficial insects, and plants. The concept — *use of a living virus to control harmful insects* — first envisioned more than a century ago, may become a reality within the next few years. Of course, many serious problems still exist. Present problems of effectiveness, safety, and feasibility were generally defined and discussed. Most assuredly, newer problems will arise as we develop and use viral insecticides. In spite of these problems, viral insecticides should find a place in the future as selective insect control agents. How much of a future, depends mainly upon its proponents and development over the next several years.

References

AIZAWA, K.: Some methods of keeping the virus activity of the silkworm jaundice. Sansi-Kenkyu **3**, 75—77 (1953a). [Japanese].

— On the inactivation of the silkworm jaundice virus. Jap. J. appl. Zool. **17**, 181—190 (1953b). [Japanese with English summary].

— Inactivation of the silkworm jaundice virus by the ultraviolet irradiation. J. Sericult. Sci. Japan **24**, 398—399 (1955). [Japanese with English summary].

— Infection of the greater wax moth, *Galleria mellonella* (Linnaeus), with the nuclear polyhedrosis virus of the silkworm, *Bombyx mori* (Linnaeus). J. Insect Path. **4**, 122—127 (1962).

ALLEN, G. E., B. G. GREGORY, and J. R. BRAZZEL: Integration of the *Heliothis* nuclear polyhedrosis virus into a biological control program on cotton. J. econ. Entomol. **59**, 1333—1336 (1966).

—, and T. L. PATE: The potential role of a feeding stimulant used in combination with the nuclear polyhedrosis virus of *Heliothis*. J. Invertebrate Path. **8**, 129—130 (1966).

ARUGA, H.: Induction of virus infections. In: Insect pathology — An advanced treatise (E. A. STEINHAUS, ed.), vol. 1, p. 499—530. New York: Academic Press. 1963.

—, and E. NAGASHIMA: Generation-to-generation transmission of the cytoplasmic polyhedrosis virus of *Bombyx mori* (Linnaeus). J. Insect Path. **4**, 313—320 (1962).

—, and H. WATANABE: Resistance to *per os* infection with cytoplasmic polyhedrosis virus in the silkworm, *Bombyx mori* (Linnaeus). J. Insect Path. **6**, 387—394 (1964).

BALCH, R. E., and F. T. BIRD: A disease of the European spruce sawfly, *Gilpinia hercyniae* (Htg.), and its place in natural control. Scient. Agricult. Ottawa **25**, 65—80 (1944).

BEN-SHAKED, Y., and I. HARPAZ: Protection of a susceptible insect host against a nuclear polyhedrosis virus by ether extracts from insect larvae. J. Invertebrate Path. **8**, 283—285 (1966).

BERGOLD, G. H.: Über Polyeder-Krankheiten bei Insekten. Biol. Zbl. **63**, 1—55 (1943).

— On the nomenclature and classification of insect viruses. Ann. N.Y. Acad. Sci. **56**, 495—516 (1953a).

— Insect viruses. Advanc. Virus Res. **1**, 91—139 (1953b).

— Viruses of insects. In: Handbuch der Virusforschung (K. F. MEYER u. C. HALLAUER, Hrsg.), vol. IV, Suppl. 3, p. 60—142. Wien: Springer 1958.

— The nature of nuclear polyhedrosis viruses. In: Insect pathology — An advanced treatise (E. A. STEINHAUS, ed.), vol. 1, p. 413—456. New York: Academic Press. 1963.

— Insect viruses. In: Techniques in experimental virology (R. J. C. HARRIS, ed.), p. 112—144. New York: Academic Press 1964.

— K. AIZAWA, K. M. SMITH, E. A. STEINHAUS, and C. VAGO: The present status of insect nomenclature and classification. Report of Insect Virus Study Group of the Subcommittee on Viruses. Intern. Bull. Bacteriol. Nomenclature and Taxonomy **10**, 259—262 (1960).

BERKALOFF, A., J. BREGLIANO, and A. OHANESSIAN: Mise en evidence de virions des drosophiles infectees par le virus hereditaire. C. R. Acad. Sci. (Paris) **260**, 5956—5959 (1965).

BILIOTTI, E.: Entomophages et maladies des insectes. Entomophaga **1**, 45—53 (1956).

— P. GRISON, and D. MARTOURET: L'utilisation d'une maladie a virus comme methode de lutte biologique contre *Pieris brassicae* L. Entomophaga **1**, 35—44 (1956).

BILIOTTI, E., F. GRISON, R. MAURY, and C. VAGO: Emploi d'une poudre a base de virus specifiquo contre la chenille processionnaire du pin dans le massif du Ventous. C. R. Acad. Agric. France **45**, 407—411 (1959a).
— —, and C. VAGO: Essai d'utilisation des polyedres isoles de la processionnaire du pin, comme methode de lutte biologique. C. R. Acad. Sci. (Paris) **243**, 206—208 (1959b).
BIRD, F. T.: The dissemination and propagation of a virus disease affecting the European pine sawfly, *Neodiprion sertifer* (Geoff.). Canada, Dominion Dept. of Agr. Forest Insect Investigations Bi-Monthly Progress Report vol. 6 (5), p. 2—3, Sept.—Oct. (1950).
— The use of a virus disease in the biological control of the European pine sawfly, *Neodiprion sertifer* (Geoff.). Canad. Entomologist **85**, 437—446 (1953).
— Virus diseases of sawflies. Canad. Entomologist **87**, 124—127 (1955).
— Transmission of some insect viruses with particular reference to ovarial transmission and its importance in the development of epizootics. J. Insect Path. **3**, 352—380 (1961).
—, and J. M. BURK: Artifically disseminated virus as a factor controlling the European spruce sawfly, *Diprion hercyniae* (Htg.), in the absence of introduced parasites. Canad. Entomologist **93**, 228—238 (1961).
—, and D. E. ELGEE: Virus disease and introduced parasites as factors controlling the European spruce sawfly, *Diprion hercyniae* (Htg.), in central New Brunswick. Canad. Entomologist **89**, 371—378 (1957).
—, and M. M. Whalen: A virus disease of the European pine sawfly, *Neodiprion sertifer* (Geoff.) Canad. Entomologist 85, 433—437 (1953).
BRAZZEL, J. R.: Potentials for combining chemical and biological insect control. 1965 Proc. Beltwide Cotton Production Mechanization Conf., p. 24—25(1965).
BURGDORFER, W., and M. G. R. VARMA: Transtadial and transovarial development of disease agents in arthropods. In: Annual review of entomology (R. F. SMITH and T. E. MITTLER, eds.). Acad. Rev. Palo Alto, Calif., **13**, 347—376 (1967).
BURGERJON, A., and P. GRISON: Adhesiveness of preparations of *Smithiavirus pityocampae* Vago on pine foliage. J. Invertebrate Path. **7**, 281—284 (1965).
CAMERON, J. W. MACB.: Factors affecting the use of microbial pathogens in insect control. In: Annual review of entomology (R. F. SMITH and T. E. MITTLER, eds.). Acad. Rev. Palo Alto, Calif., **8**, 265—286 (1963).
CANTWELL, G. E., A. K. DAVID, L. THOR, and A. S. MICHAEL: Mortality of the honey bee, *Apis mellifera*, in colonies treated with certain biological insecticides. J. Invertebrate Path. **8**, 228—233 (1966).
CARLETON, W. M., and R. D. BRAZEE: Role of agricultural engineers in pest control by chemical, biological, genetic and physical means. In: U.S.D.A. Bulletin ARS 33-110, 214 pages (1966).
CHANT, D. A.: Strategy and tactics of insect control. Canad. Entomologist **96**, 182—201 (1964).
CHAPMAN, A. T., R. A. BELL, and C. M. IGNOFFO: Field tests with the nuclear polyhedrosis virus for control of *Heliothis* in the southern high plains of Texas. J. econ. Entomol. (in press).
CLARK, E. C.: Observations on the ecology of a polyhedrosis of the Great Basin tent caterpillar, *Malacosoma fragilis*. Ecology **36**, 373—376 (1955).
—, and O. E. REINER: The availability of certain proprietary adjuvants for use with the polyhedrosis viruses of insects. J. econ. Entomol. **49**, 703—704 (1956).
—, and C. G. THOMPSON: The possible use of microorganisms in the control of the Great Basin tent caterpillar. J. econ. Entomol. **47**, 268—272 (1954).
CORNALIA, E.: Monografia del bombice del gelso. Mem. R. Instit. Lambardo Sci. Lett. Arte **6**, 3—387 (1856).
DARLINGTON, C. D.: The evaluation of genetic systems. New York: Basic Books 1958.

DAVID, W. A. L., and B. O. C. GARDINER: A *Pieris brassicae* (Linnaeus) culture resistant to a granulosis. J. Insect Path. **2**, 106—114 (1960).
— — The incidence of granulosis deaths in susceptible and resistant *Pieris brassicae* (Linnaeus) larvae following changes of population density, food, and temperature. J. Invertebrate Path. **7**, 347—355 (1965a).
— — Resistance of *Pieris brassicae* (Linnaeus) to granulosis virus and the virulence of the virus from different host races. J. Invertebrate Path. **7**, 285—290 (1965b).
— — Persistence of a granulosis virus of *Pieris brassicae* on cabbage leaves. J. Invertebrate Path. **8**, 180—183 (1966a).
— — Breeding *Pieris brassicae* apparently free from granulosis virus. J. Invertebrate Path. **8**, 325—333 (1966b).
DUBOS, R.: Pasteur and modern science. New York: Anchor Books 1960.
ELMORE, J. C.: Control of the cabbage looper with a nuclear polyhedrosis virus disease. J. econ. Entomol. **54**, 47—50 (1961).
—, and A. F. HOWLAND: Natural versus artifical dissemination of nuclear polyhedrosis virus by contaminated adult cabbage loopers. J. Insect Path. **6**, 430—438 (1964).
FALCON, L.: Personal communication (1967).
FALCON, L. A., T. F. LEIGH, R. VAN DEN BOSCH, J. H. BLACK and V. E. BURTON: Insect diseases tested for control of cotton bollworm. Calif. Agricult. **19**, 13—14 (1965).
FRANZ, J. M.: Biological control of pest insects in Europe. Ann. Rev. Entomol. **6**, 183—200 (1961).
GENUNG, W. G.: Comparison of insecticides, insect pathogens and insecticide-pathogen combinations for control of cabbage looper, *Trichoplusia ni* (HÜBNER). Florida Entomologist **43**, 65—68 (1960).
GERSHENSON, S.: On the species specificity of viruses of the polyhedral diseases of insects. Mikrobiologia **24**, 90—98 (1955) [Russian].
GETZIN, L. W.: The effectiveness of the polyhedrosis virus for control of the cabbage looper, *Trichoplusia ni*. J. econ. Entomol. **55**, 442—445 (1962).
GILMORE, J. E.: Preliminary field evaluation of a noninclusion virus for control of the citrus red mite. J. econ. Entomol. **58**, 1136—1140 (1965).
—, and H. TASHIRO: Fecundity, longevity, and transinfectivity of citrus red mites (*Panonychus citri*) infected with a noninclusion virus J. Invertebrate Path. **8**, 334—339 (1966).
GLASS, E. H.: Laboratory and field tests with the granulosis of the redbanded leaf roller. J. econ. Entomol. **51**, 454—457 (1958).
GRISON, P.: Utilisation en foret d'une preparation a base de virus specifique contre *Thaumatopoea pitvocampa* Schiff. Sonderdruck aus Z. angew. Entomol. **47**/1, S. 24—31 (1960).
— Personal communication. (1967).
—, C. VAGO, and R. MAURY: La lutte contre la processionnaire du pin "*Thaumetopoea pitycampa*" Schiff dans le massif du Ventoux; essai d'utilisation practique d'un virus specifique. Rev. Forestiere Franc. **5**, 353—370 (1959).
HARPAZ, I., and Y. BEN-SHAKED: Generation-to-generation transmission of the nuclear polyhedrosis virus of *Prodenia litura* (Fabricius). J. Insect Path. **6**, 127—130 (1964).
—, and E. ZLOTKIN: A nuclear polyhedrosis virus of the safflower leaf worm *Heliothis peltigera* Schiff. (Lep., Noctudiae). Ann. Soc. Ent. Fr. (N.S.) **1** (4), 963—972 (1965).
HEIMPEL, A. M.: Microbial control of insects. Wld Rev. Pest Control **4**, 150—159 (1965).
— Exposure of white mice and guinea pigs to the nuclear polyhedrosis virus of the cabbage looper, *Trichoplusia ni*. J. Invertebrate Path. **8**, 98—102 (1966).

HEIMPEL, A. M., and L. K. BUCHANAN: Human feeding tests using the nuclear polyhedrosis virus. J. Invertebrate Path. **9**, 55—57 (1967).
HIRT, R. C., R. G. SCHMITT, N. D. SEARLE, and A. P. SULLIVAN: Ultraviolet spectral energy distributions of natural sunlight and accelerated test light sources. J. opt. Soc. Amer. **50**, 706—713 (1960).
HOFMASTER, R. N., and L. P. DITMAN: Utilization of a nuclear polyhedrosis virus to control the cabbage looper on cole crops in Virginia. J. econ. Entomol. **54**, 921—923 (1961).
HOLMES, F. O.: Order virales, the filterable viruses. In: BERGEY'S Manual of determinative bacteriology, 6th ed., p. 1225—1228. Baltimore: Williams & Wilkins Co. 1948.
HUGER, A.: Granulosis of insects. In: Insect pathology — An advanced treatise (E. A. STEINHAUS, ed.), vol. 1, 531—575. New York: Academic Press 1963.
HUKUHARA, T.: Generation-to-generation transmission of the cytoplasmic polyhedrosis virus of the silkworm, *Bombyx mori* (Linnaeus). J. Insect Path. **4**, 132—135 (1962).
IGNOFFO, C. M.: Production and virulence of a nuclear polyhedrosis virus from larvae of *Trichoplusia ni* (HÜBNER) reared on a semisynthetic diet. J. Insect Path. **6**, 318—326 (1964).
— The nuclear polyhedrosis virus of *Heliothis zea* (Boddie) and *Heliothis virescens* (FABRICIUS). IV. Bioassay of virus activity. J. Invertebrate Path. **7**, 315—319 (1965a).
— The nuclear polyhedrosis virus of *Heliothis zea* (BODDIE) and *Heliothis virescens* (F.). Part 1. Virus propagation and its virulence. J. Invertebrate Path. **7**, 209—216 (1965b).
— The nuclear polyhedrosis virus of *Heliothis zea* (BODDIE) and *Heliothis virescens* (F.). Part II. Biology and propagation of diet reared *Heliothis*. J. Invertebrate Path. **7**, 217—226 (1965c).
— Evaluation of the polyhedrosis virus for bollworm control. 1956 Proc. Beltwide Cotton Production Mechanization Conf., p. 26—27 (1965d).
— Production, identification, and standardization of insect viral pathogens. Entomophaga **10** (1), 29—40 (1965e).
— Insect viruses. In: Insect colonization and mass production (C. M. SMITH, ed.), chapter 36, p. 501—530. New York: Academic Press 1966a.
— Susceptibility of the first instar of the bollworm, *Heliothis zea*, and the tobacco budworm, *Heliothis virescens*, to *Heliothis* nuclear polyhedrosis virus. J. Invertebrate Path. **8**, 531—536 (1966b).
— Effects of temperature on mortality of *Heliothis zea* larvae exposed to sublethal doses of a nuclear polyhedrosis virus. J. Invertebrate Path. **8**, 290—292 (1966c).
— Effects of age on mortality of *Heliothis zea* and *Heliothis virescens* larvae exposed to a nuclear polyhedrosis virus. J. Invertebrate Path. **8**, 279—282 (1966d).
— Possibilities of mass-producing insect pathogens. Internat. Colloq. Insect Pathol. Netherlands, Sept. 1966 (in press) (1967).
—, and J. R. ADAMS: A cytoplasmic polyhedrosis virus, *Smithiavirus pectinophorae* sp. n. of the pink bollworm, *Pectinophora gossypiella* (SAUNDERS). J. Invertebrate Path. **8**, 59—66 (1966).
— A. J. CHAPMAN, and D. F. MARTIN: The nuclear polyhedrosis virus of *Heliothis zea* (BODDIE) and *Heliothis virescens* (F.). III. Effectiveness of the virus against field populations of *Heliothis* on cotton, corn, and grain sorghum. J. Invertebrate Path. **7**, 227—235 (1965).
—, and C. GARCIA: The relation of pH to the activity of inclusion bodies of a *Heliothis* nuclear polyhedrosis. J. Invertebrate Path. **8**, 426—427 (1966).
—, and A. M. HEIMPEL: The nuclear polyhedrosis virus of *Heliothis zea* (BODDIE) and *Heliothis virescens* (F.). V. Toxicity-pathogenicity of virus to white mice and guinea pigs. J. Invertebrate Path. **7**, 329—340 (1965).

IGNOFFO, C. M., and E. L. MONTOYA: The effects of chemical insecticides and insecticidal adjuvants of a *Heliothis* nuclear polyhedrosis virus. J. Invertebrate Path. **8**, 409—412 (1966).
JAQUES, R. P.: The influence of physical stress on growth and nuclear polyhedrosis of *Trichoplusia ni* (HÜBNER). J. Insect Path. **3**, 47—54 (1961).
— Stress and nuclear polyhedrosis in crowded populations of *Trichoplusia ni* (HÜBNER). J. Insect Path. **4**, 1—22 (1962a).
— The transmission of nuclear polyhedrosis virus in laboratory populations of *Trichoplusia ni* (HÜBNER). J. Insect Path. **4**, 433—445 (1962b).
— The persistence of a nuclear polyhedrosis virus in soil. J. Insect Path. **6**, 251—254 (1964).
KELSEY, J. M.: Virus spray for control of *Pieris rapae* L. New Zealand J. Sci. Technol. Sect. A **38**, 644—646 (1957).
— Control of *Pieris rapae* by granulosis virus. New Zealand J. Agri. Res. **1**, 778—782 (1958).
KRIEG, A.: Untersuchungen zur Wirbeltier-Pathogenität und zum serologischen Nachweis der *Rickettsia melolonthae* im Arthropod-Wirt. Naturwissenschaften **42**, 609—610 (1955).
LEIGH, T. F., J. H. BLACK, C. E. JACKSON, and V. E. BURTON: Insecticides and beneficial insects in cotton fields. Calif. Agriculture **20**, 4—6 (1966).
L'HERITIER, P.: Sensitivity to CO_2 in Drosophila — A review. Heredity **2**, 325—348 (1948) .
LEWIS, F. B.: How to collect and process small polyhedral viruses of insects. U.S. Dep. Agric. Forest Research Notes, Northeastern Forest Exp. Sta. No **109**, 1—8 (1960).
LWOFF, A., and P. TOURNIER: The classification of viruses. In: Annual review of microbiology (eds. C. E. CLIFTON, S. RAFFEL and M. P. STARR), vol. **20**, p. 45—74 1966.
MALESTRI, A.: Frammenti anatomici fisiologici et patologici sue baso da seta, 172 p. Pavea: Fuse 1856.
MARTIGNONI, M. E.: Progressive nucleopolyhedrosis in adults of *Peridroma saucia* (HÜBNER). J. Insect Path. **6**, 368—372 (1964a).
— Mass production of insect pathogens. In: Biological control of insect pests and weeds (P. H. DE BACH, ed.), p. 579—609. New York: Reinhold Publ. Co. 1964b.
—, and J. E. MILSTEAD: Trans-ovum transmission of the nuclear polyhedrosis virus of *Colias eurytheme* Boisduval through contamination of the female genitalia. J. Insect Path. **4**, 113—121 (1962).
—, and P. SCHMID: Studies on the resistance to virus infections in natural populations of Lepidoptera. J. Insect Path. **3**, 62—74 (1961).
MARTOURET, D., and G. DUSAUSSOY: Multiplication et extraction des corps d'inclusion de la virose intestinale de *Thaumetopoea pityocampa* Schiff. Entomophaga **4**, 253—259 (1959).
—, R. MAURY, and C. VAGO: Essais d'utilisation de suspension de polyedres contre *Thaumetopoea pityocampa* Schiff. Extrait des Comptes redus du IV Congres International de Lutte contre des Ennemis des Plantes Hambourg 1957, vol. 1, p. 889—890. Braunschweig 1959.
MCEWEN, F. L., and G. E. R. HERVEY: Control of the cabbage looper with a virus disease. J. econ. Entomol. **51**, 626—631 (1958).
— — Microbial control of two cabbage insects. J. Insect Path. **1**, 86—94 (1959).
MCLAUGHLIN, R. E.: Infection of the boll weevil with *Mattesia grandis* induced by a feeding stimulant. J. econ. Entomol. **59**, 909—910 (1966).
MONTOYA, E. L., C. M. IGNOFFO, and R. L. MCGARR: A feeding stimulant to increase effectiveness of, and a field test with, a nuclear polyhedrosis virus of Heliothis. J. Invertebrate Path. **8**, 320—324 (1966).

NEILSON, M. M.: A cytoplasmic polyhedrosis virus pathogenic for a number of lepidopterous hosts. J. Insect Path. **6**, 41—52 (1964).
— Effects of a cytoplasmic polyhedrosis on adult lepidoptera. J. Invertebrate Path. **7**, 306—314 (1965).
—, and D. E. ELGEE: The effect of storage on the virulence of a polyhedrosis virus. J. Insect Path. **2**, 165—171 (1960).
OSSOWSKI, L. L. J.: The biological control of the wattle bagworm (*Kotochalia junodi* Heyl.) by a virus disease. I. Small-scale pilot experiments. Ann. appl. Biol. **45**, 81—89 (1957).
— Occurrence of strains of the nuclear polyhedral virus of the wattle bagworm. Nature (Lond.) **181**, 648 (1958).
— Variation in virulence of a wattle bagworm virus. J. Insect Path. **2**, 35—43 (1960a).
— The biological control of the wattle bagworm, *Kotochalia junodi* (Heyl.), by a virus disease. II. Large-scale experiments. Ann. appl. Biol. **48**, 299—313 (1960b).
PAILLOT, A.: Sur une nouvelle maladie des chenilles de *Pieris brassicae* et sur les maladies du noyau chez les insectes. C. R. Acad. Sci. (Paris) **179**, 1353—1356 (1924).
— Sur une nouvelle maladie du noyau ou grasserie des chenilles de *P. brassicae* et un naiveau groupe de microorganismes parasites. C. R. Acad. Sci. (Paris) **182**, 180—182 (1926).
POLTEV, V. I.: Microbiological control method in beekeeping. Proc. Intern. Beekeeping Congr., Moscow, vol. 19, p. 224—249 [in Russian] (1963).
PONSEN, M. B.: The production of polyhedral viruses from *Barathra brassicae* to control *Adoxophyes reticulana*. Overdruk Uit: Mededelingen Rijksfaculteit Landbouw. Wetenschappen Gent, 1966, XXXI, Nr. 3 (1966).
RIVERS, C. F.: Virus resistance in larvae of *Pieris brassicae* (L.) Trans. 1st Intern. Conf. Insect Pathol. and Biol. Control, Praha 1958, p. 205—210. 1959.
—, and D. GLITZ: Personal communication (1967).
SAGER, S. M.: On the transtadial transmission of insect viruses. J. Insect Path. **2**, 307—309 (1960).
SCHMIDT, L., and G. PHILIPS: Granulosis — a new virus disease of the fall webworm, *Hyphantria cunea* Drury. Poljoprivredno Sumarski Fakultet, Zavod. za entomologiju, Zagreb **1**, 1—27 [Orig. Jugoslawisch] (1958).
SHAW, J. C.: Poison-lure sprays for the Mexican fruit fly. Calif. Citrograph **40**, 188—192 (1955).
SMIRNOFF, W. A.: A virus disease of *Neodiprion swainei* Middleton. J. Insect Path. **3**, 29—46 (1961).
— Trans-ovum transmission of virus of *Neodiprion swainei* Middleton (Hymenoptera, Tenthredinidae). J. Insect Path. **4**, 192—200 (1962).
— Preparation and application of viral material, in biological control of the jack pine sawfly. Forestry Chronicle **40**, 187—194 (1964).
—, and C. F. MACLEOD: Apparent lack of effects of orally introduced polyhedrosis virus on mice and of pathogenicity of rodentpassed virus for insects. J. Insect Path. **6**, 537—547 (1964).
SMITH, K. M.: The cytoplasmic virus diseases. In: Insect pathology: An advanced treatise (E. A. STEINHAUS, ed.). New York: Academic Press 1963.
— G. H. HILLS, and C. F. RIVERS: Studies on the cross-inoculation of the Tipula irridescent virus. Virology **13**, 233—241 (1961).
—, and R. C. WILLIAMS: Insect viruses and their structure, Endeavour vol. **17**, 12—21. 1958.
—, and N. Xeros: Studies on the cross-transmission of polyhedral viruses: Experiments with a new virus from *Pyrameis cardui*, the painted lady butterfly. Parasitology **43**, 178—185 (1953).

SMITH, O. J., K. M. HUGHES, P. H. DUNN, and I. M. HALL: A granulosis virus disease of Western grape leaf skeltonizer and its transmission. Canad. Entomologist **88**, 507—515 (1956).
STAIRS, G. R.: Quantitative differences in susceptibility to nuclear polyhedrosis virus among larval instars of the forest tent caterpillar, *Malacosoma disstria* (HÜBNER). J. Invertebrate Path. **7**, 427—429 (1965).
STEINHAUS, E. A.: Principles of insect pathology, 757 p. New York: McGraw-Hill Book Co. 1949.
— Taxonomy of insect viruses. Ann. N. Y. Acad. Sci. **56**, 517—537 (1953).
— Stress as a factor in insect disease. Proc. Intern. Congr. Entomol. 10th Montreal, vol. 4, p. 725—730 (1956).
— Crowding as a possible stress factor in insect disease. Ecology **39**, 503—514 (1958).
— The duration of viability and infectivity of certain insect pathogens. J. insect Path. **2**, 225—229 (1960a).
— Notes on polyhedroses in *Peridroma*, *Prodenia*, *Colias*, *Heliothis*, and other *Lepidoptera*. J. Insect Path. **2**, 327—333 (1960b).
—, and J. P. DINEEN: Observations of the role of stress in a granulosis of the variegated cutworm. J. Insect Path. **2**, 55—65 (1960).
—, and C. G. THOMPSON: Preliminary field tests using a polyhedral virus to control the alfalfa caterpillar. J. econ. Entomol. **42**, 301—305 (1949).
TANADA, Y.: Description and characteristics of a granulosis virus of the imported cabbageworm. Proc. Hawaiian Entomol. Soc. **15**, 235—260 (1953).
— Microbial control of imported cabbageworm. Hawaii Farm Sci. **4**, 6—7 (1956a).
— Microbial control of some lepidopterous pests of crucifers. J. econ. Entomol. **49**, 320—329 (1956b).
— Microbial control of insect pests. Ann. Rev. Entomol. **4**, 277—302 (1959).
— A granulosis virus of the codling moth, *Carpocapsa pomonella* (Linnaeus) (Olethreutidae, Lepidoptera). J. Insect Path. **6**, 378—380 (1964).
— Personal communication (1967).
—, and C. REINER: The use of pathogens in the control of the corn earworm, *Heliothis zea* (BODDIE). J. Insect Path. **4**, 139—154 (1962).
—, and A. M. TANABE: Resistance of *Galleria mellonella* (Linnaeus) to the *Tipula* iridescent virus at high temperatures. J. Invertebrate Path. **7**, 184—188 (1965).
THOMPSON, A. D., and B. A. SMIRK: Rod-shaped particles associated with a disease of alkali bees. Virology **28**, 348—349 (1966).
THOMPSON, C. G.: Thermal inhibition of certain polyhedrosis virus diseases. J. Insect Path. **1**, 189—192 (1959).
—, and E. A. STEINHAUS: Further tests using a polyhedrosis virus to control the alfalfa caterpillar. Hilgardia **19**, 411—445 (1950).
TURNER, G. S., and C. KAPLAN: Observations on photo-dynamic inactivation of vaccinia virus and its effect on immunogenicity. J. Hyg. Camb. **63**, 395 (1965).
VAGO, C.: Multiplication du virus *Borrelina bombycis* sur chrysalides de vers a soie. C. R. Acad. Sci. (Paris) **245**, 2115—2117 (1957).
— Sur la nomenclature des virus d'insectes. Entomophaga **3**, 331—332 (1958).
— Sur la potentialites des germes entomopathogenes an lutte microbiologique. Rev. Path. Veg. et Ent. **43**, 101—111 (1964).
—, and P. ATGER: Multiplication massive des virus d'insectes pendant la mue nymphale. Entomophaga **6**, 53—56 (1961).
—, and R. CAYROL: Une virose a polyedres de la noctuella gamma *Plusia gamma* L. (Lepidoptera). Ann. Inst. Natl. Recherches Agron. Ser. C. Ann. Epiphyt. **6**, 421—432 (1955).
VAIL, P.: Personal communication (1967).
WALLIS, R. C.: Incidence of polyhedrosis of gypsy moth larvae and the influence of relative humidity. J. econ. Entomol. **50**, 580—583 (1957).

WASSER, H. B.: Demonstration of a new insect virus not associated with inclusion bodies. J. Bact. **64** (1952).
WATANABE, H.: Resistance to peroral infection by the cytoplasmic polyhedrosis virus in the silkworm, *Bombyx mori* (Linnaeus). J. Invertebrate Path. **7**, 257—270 (1965).
WATANABE, S.: Studies on the grasserie virus of the silkworm, *Bombyx mori*. IV. Physical and chemical effects upon the virus. Japan. J. exp. Med. **21**, 299—313 (1951).
WEISER, J.: Zur Taxonomie der Insektenviren. Cs. Parasit. **5**, 203—211 (1958).
— Vagoiavirus gen. n., a virus causing disease in insects. J. Invertebrate Path. **7**, 82—85 (1965).
—, and C. VAGO: A newly described virus of the winter moth, *Operophthera brumata* HÜBNER (Lepidoptera, Geometridae). J. Invertebrate Path. **8**, 314—319 (1966).
WOLFENBARGER, D. A.: Paraffinic and naphthenic oil fractions in combinations with DDT and a *Heliothis* virus for corn earworm control. J. econ. Entomol. **57**, 732—735 (1964).
— Polyhedrosis-virus-surfactant and insecticide combinations, and *Bacillus thuringiensis*-surfactant combinations for cabbage looper control. J. Invertebrate Path. **7**, 33—38 (1965).
ZHDANOV, V. M.: Guide to human and animal viruses. Izdatelstvo Academii Medicinskich Nauk U.S.S.R. Moscow, p. 316—323 (348 p.), Publ. Acad. Med. Sci. U.S.S.R. Moscow (1953).

Subject Index

Acarina 27, 140, 142
—, virosis of 27, 140, 142
Acrobasi zelleri 138
acute safety test of viral insecticides in vertebrate animals 153
adaptation of nuclear-polyhedrosis viruses to unnatural hosts 15
adsorption of nuclear polyhedrosis virus to cells in suspension 117
— — — — — —, factors influencing this event 117
Aedes aegypti as transmitter of vesicular stomatitis virus 43
— —, cell line from 109
— —, primary embryonic cell culture 108, 110
— — larvae, multiplication of Egypt 101 West Nile virus in primary cultures of immaginal discs from 123
— — tissue cultures, survival and replication of western equine encephalomyelitis virus in 123
African horse sickness 43
Agallia constricta, distribution of wound tumor virus in central nervous system of 98
— —, multiplication of wound tumor virus in 95
— — primary tissue cultures, infection with wound tumor virus 122
— —, sites of virus within tissue and cells of organs of 96
— —, vector and host of wound tumor virus 96
Agalliopsis novella, inefficiency as vector of wound tumor virus 98
— —, infection with wound tumor virus 98
agricultural compatibility of viruses 148
allergenicity of viral insecticides 153
American foulbrood of the bee 26
amino acids in hemolymph, decrease during cytoplasmic polyhedroses 10
— —, level in hemolymph 110
— — of granulosis virus protein 5
amino acids of nuclear and cytoplasmic polyhedral protein 5, 6
Antheraea cell line, growth of subline with bovine sera and egg ultrafiltrate 109
— *eucalypti*, cell line from 109
— — tissue culture, spontaneous appearance of cytoplasmic polyhedrosis virus in 114
— —, flaccidity of 29
— —, infection of cell cultures with iridescent viruses 28, 114, 158
antibody response after infection with arboviruses 47
antigenic grouping of arboviruses 40, 41
— relationship of arboviruses used for classification 39
antigenicity of nuclear-polyhedrosis and granulosis viruses 6
aphid, systemic infection by a plant pathogenic virus 104
— vectors, interaction with plant pathogenic viruses 102
aphid-borne viruses 102
— —, electron microscopic localization in plants and insects in situ 103
— —, similarities and differences to other groups of arthropod-borne viruses 102
application of arthropod viruses as living insecticides 143
arbovirus catalogue 39
arboviruses 38, 122
—, arthropod vectors 42, 49
—, Bunyamwera group 39, 40, 41, 42
—, characteristics 38
—, chemical inactivation 42
—, classification 38, 39
—, course of infection in vertebrates 47
—, criteria for definition 41, 122
—, cultivation in arthropod tissue cultures 122
—, cycle of infection 44
—, detection systems in the laboratory 50
—, development 38

arboviruses, disease control 53
—, group A 39, 40, 42
—, — A and B, multiplication and survival tested in *Hyalomma dromedarii* tissue culture 125
—, — B 39, 40, 42
—, — C 39, 40
—, Guama group 39, 40
—, having no arthropod vectors 50
—, host spectra 46
—, increase of titer in tick tissue cultures 124
—, insects as hosts and vectors 122
—, isolation from blood-suckling arthropods 41
—, — cases of disease under certain epidemiological circumstances 42
—, — intracerebrally infected suckling mice 50
—, — salivary glands of mosquitoes 42
—, — sentinel animals 42
—, — tissue culture 50, 51, 122, 125
—, multiplication in insects after artificial infection 49
—, origin 49
—, pathogenicity and symptomatology in vertebrates 49
—, pathogenicity for suckling mice 39, 42
—, public health importance 51
—, reintroduction by migrating birds 48
—, relationship to insect hosts, use of insect tissue culture for study of 123
—, reservoirs in vertebrates 48
—, secondarily vector borne 50
—, shape 38
—, size 38, 42
—, transmission 38, 43
—, ungrouped 39
—, vaccination 53
—, vertebrate hosts 46, 48, 49, 50
Arctiidae 17
Argyrotaenia velutinana 4, 144, 148
armyworms viruses 144, 159
arthropod-borne animal viruses 38, 42, 43, 44, 50, 122
arthropod pests, viruses isolated from 141
— tissue cultures, cultivation of arboviruses in 122
— vectors, control of 33
— — of arboviruses 42, 49
arthropod viruses as control agents 143
— —, classification of 134, 142
— —, structure of 1, 130
— —, types of 1, 24, 129, 134
arthropods as arbovirus reservoirs 45
artificial infection of insects with arboviruses 49
aspects of comparative virology 29
assay for sigma virus 63
— — —, based on end-point dilution 63
— methods for multiplication of plant viruses in insects 95
— system for SIV in *Antheraea* cell line 115
association of sigma virus with a cell organelle 88
aster yellows virus, multiplication in aster leafhopper tissue in vitro 121
— — —, multiplication in the vector *Macrosteles fascifrons* 95
atypical weakening of the bee 34
Austrocaligula eucalypti, cell line from 109
availability of living viral insecticides 129

***Bacillus** alvei* 26
— *euridice* 26
— *larvae* 26
— *pluton* 26
— *thuringiensis* 26
bacteria, interference with inclusion-type insect virus multiplication 17
baits, used to increase the effectiveness of field applied virus 147
Barathra brassicae 156
bat, salivary gland viruses 39
bats as arbovirus reservoirs 48
bee, atypical weakening of the 34
—, black disease of the 34
— larvae 25
—, paralysis of the 34
beet yellow net virus 102
— yellows virus 102
behavioral habits of the target pest as factor of effectiveness of viruses 147
Bergoldia viruses 134, 142
BHK-21 baby hamster kidney cells, use for isolation of arboviruses 51
biological control of harmful insect populations 33
— interactions between plant viruses and insect vectors 94

BIOTROL-VHZ 157
Birdia virus 134
birds as hosts of encephalitogenic viruses 46
—, infected with St. Louis encephalitis virus 49
black disease of the bee 34
Blattella germanica, explant culture of tissues 110
blood cells 109, 116
— — of *Chilo suppressalis*, long-term cultivation 109
— — of *Peridroma saucia*, monolayer culture of 109
— plasma depletion during nuclear-polyhedroses 10
blue tongue disease of sheep 43, 44
body fluids, increase during polyhedroses 10
bollworm-budworm complex of viruses 141, 159
— nuclear polyhedrosis 16, 156
— — —, safety to various organisms 154
Bombycidae 17
Bombyx mori, course of infection with nuclear polyhedrosis virus 8
— —, genetic constitution and response to cytoplasmic polyhedrosis virus 15
— —, infection of cell cultures with densonucleosis virus 27
— — pupae, culture of ovaries 108, 111
Borrelina viruses 134, 142
brain 105
B-vitamin complex supplement in medium for insect tissue cultures 110

cabbage looper, susceptibility to nuclear polyhedrosis virus 15, 144
— — viruses 15, 144, 159
Cache Valley virus 39
calcium ions, level in hemolymph 110
calf serum as medium supplement for insect tissue cultures 110
— — fraction as supplement for insect tissue culture medium 110
— —, substitute for hemolymph in insect tissue culture medium 110
California encephalitis virus 49
Callosamia promethea 111
Calomys callosus as vertebrate host of Machupo virus 54
carcinogenicity of viral insecticides 153
Carcinus maenas, virus infection of 26
Carpocapsa pomonella 4, 148
— —, granulosis virus, transmission to *Grapholitha molesta* 152
cell cultures of insect tissues 108
— — — —, infection with arboviruses 122
— — — —, — cytoplasmic polyhedrosis virus 113, 158
— — — —, — densonucleosis virus 27
— — — —, — granulosis virus 114
— — — —, — iridescent virus 28, 114, 158
— — — —, — nuclear polyhedrosis virus 111, 158
— — — —, — plant viruses 121
cellular resistance of embryonic cells to nuclear-polyhedrosis viruses 11
— — to sigma virus, possible explanations 86, 88
— — superinfection of sigma virus 86, 88
central nervous system, distribution of wound tumor virus in 98
characteristics of arboviruses 38
— the major arthropod viruses 135
CHASTANG's medium 110
chemical analysis of rod-shaped virions of inclusion-type insect viruses 6
— composition of non-inclusion type insect viruses 25, 27
chemicals inducing development of resistance against viruses 151
chemistry of inclusion-type insect viruses 1
chick embryo cell cultures, used for isolation of arboviruses 51
— — extract as medium supplement for insect tissue cultures 110
Chikingunya virus 49
Chilo iridescent virus, development in primary cultures of hemocytes, fat body cells, gonadal tissue and brain of *Chilo suppressalis* 115
— *suppressalis* cell cultures, development of purified iridescent virus in 115
— —, infection with iridescent virosis 28

Chilo suppressalis, long-term cultivation of blood cells 109
Choristoneura fumiferana 4, 17
— —, interference of granulosis and nuclear-polyhedrosis viruses in 17
chromosome site for sigma virus 88
chronic safety test of viral insecticides in vertebrate animals 153
circulative aphid-borne virus, electron microscopic localization in plants and insects in situ 103
Cirphis unipuncta, viral disease of 25
classification of arboviruses 38, 39, 40, 41, 42
— arthropod viruses 134, 142
— non-inclusion type viruses of invertebrates 31, 134
clear heads disease of silkworms 26
clover-infecting virus, multiplication in leafhopper vectors 95
"clusters" of viruses in non-inclusion type viroses 35
CNS-symptoms, produced by arboviruses 49
Cocal virus 43
codling moth 3
— — granulosis 3, 156
— — —, safety to various organisms 154
Coleoptera 142, 152
Colias eurytheme 147, 152
Colladonus montanus, cytopathogenic and systemic effects caused by Western X-disease virus on 96
Colorado tick fever virus 43, 49
— — — — in wild rodent hosts 49
comparative virology, aspects of 29
complement fixation test, use for classification of arboviruses 39
completion of nuclear polyhedrosis virus development in vitro in explants from diseased silkworm 112
composition of arthropod viruses 130
— insect hemolymph 110
concentration of sigma virus 65
— suspensions of inclusion bodies, determination 13
continuous insect cell lines 109
control agents, arthropod viruses as 143
— of arbovirus diseases 53
— arthropod vectors 33
CO_2-sensitivity as assay for sigma 63
— in rho-carrying flies 71
CO_2-sensitivity in ultra-rho carrying flies 71
—, distribution in Poisson fashion 63
—, influence of temperature on 72
—, relation to sigma virus titer 63
— of *D. melanogaster*, cytoplasmic agent (genoid) responsible for 60
— — — infected with sigma virus 59
— — — — — —, explanation 59
— — —, infectious nature of the responsible sigma agent 60
— — —, inheritance patterns 60
— — —, transmission by egg and sperm 60
— — —, — inoculation of material of affected animals 60
— sigma virus infected *Drosophila melanogaster* 59, 61, 63
CO_2-resistance, dependance on fly genotype 84
—, — relationships among virus types and fly genomes 85
cost to develop new insecticides 155
course of infection with arboviruses in vertebrates 47
coverage of plants with virus 146
crane fly, non inclusion virus, experimental transmission to *Diptera*, *Lepidoptera*, and *Coleoptera* 152
criteria used for classification of arthropod viruses 134, 142
cuboidal-type inclusions 4
cuboidal variant of granulosis virus in *Choristoneura fumiferana* 5
Culex pipiens quinquefasciatus as vector species of St. Louis encephalitis virus 54
Culicoides as vectors of arboviruses 43, 44, 54
culture of explants of insect adult tissue 110
— insect embryonic tissue, difficulties 108
— — tissue 108
cycle of infection of arboviruses 44
cytopathic changes in cells infected by cytoplasmic polyhedrosis virus 9
— — — — granulosis virus 9
— — — — nuclear polyhedrosis virus 8, 112, 118
— — fat body cell cytoplasm of aster leafhoppers confined to virus-infected plants 95

cytopathic changes in insect ovarian tissue cultures infected with nuclear polyhedrosis virus 112
cytopathogenic and systemic changes in leafhopper caused by Western X-disease virus 96
— effect of wound tumor virus on the nerve cells of the insect host 98
cytoplasmic inclusion bodies 1
— matrix, viroplasms formed within 101
— polyhedra, chemical studies 6
— polyhedroses 1, 133
— polyhedrosis of the pine processionary caterpillar, industrial production of the virus of 156
— — salt marsh caterpillar, *Estigmene acrea* 4
— — viruses 2, 113, 134
— — —, course of infection 9
— — —, cytopathic effect in infected cells 9
— — —, development in insect tissue cultures 114
— — —, — primary cultures of ovarian tissue 116
— — —, induction of a latent infection in *Antheraea eucalypti* tissue culture 114
— — —, release and concentration of virions from inclusion bodies 14
— — —, response of *Bombyx mori* depending on genetic constitution of host strains 15
— — —, sites of development in vivo 116
— — —, spherical shape and RNA content of 113

"décrochage" 77
—, frequency of 78
—, time of the event in oogenesis 78
deleterious effects of plant pathogenic viruses in their insect hosts 95
dengue virus complex of arbovirus group B 41, 52
density-dependent control agents 145
— independent control agents 145
Densonucleosis 27, 33
—, spectrum of virulence 27, 33
deoxycytidine, effect on sigma virus multiplication 65
—, reversal of 5-FDU and thymidine blocks of sigma virus multiplication by 65
Dermacentor marginatus, explant culture of adult tissue 110
— —, multiplication of TE virus, strain Hypr in tissue cultures from 124
— *pictus*, explant culture of adult tissue 110
— — multiplication of EEE virus in cultures of fat body and hypodermis tissue of 124
detection systems for arboviruses in the laboratory 50
determination of inclusion body concentration 13
development of arboviruses 38
— germ cells in *Drosophila* 61
— insect viruses in vitro 117
— non-inclusion type viruses of invertebrates in the cell 27, 28, 30
— wound tumor virus in insect vectors, sequential stages 102
dextran as supplement for insect tissue culture medium 110
diagnostic differentiation between nuclear and cytoplasmic polyhedrosis 137
diameter of inclusion bodies 1, 135
differences in nucleic acid content of nuclear-polyhedrosis and granulosis virions 6
different ways of infection of host insects with inclusion-type insect viruses 8
differentiation of host cells and susceptibility to nuclear-polyhedrosis viruses 11
"dimensional type" of sigma virus strains 70
dimensions of non-inclusion type insect viruses 25, 26, 27, 28, 29, 135
— sacbrood virus of the bee 25
— the virus, infecting *Cirphis unipuncta* 25
Diprion 15, 16, 144, 147, 150, 151
Diptera 16, 59, 140, 142, 151
—, inclusion-type viruses of 16, 140
—, non-inclusion type viruses of 32, 140
disease-free host populations for quantitative tests on insect viruses 13
diseases of insects, use for selective control 129

dispersion of insect ovarian and epidermal tissue 109
— nuclear-polyhedrosis viruses 19
dissemination of inclusion-type insect viruses 13
distribution of viruses among arthropods 140
DNA content of rod-shaped, membrane surrounded virions of nuclear-polyhedrosis and granulosis 6, 135
—, increase in midgut cells during cytoplasmic polyhedroses 10
— -non-inclusion viruses of invertebrates 27, 28, 32, 135
— of tissue culture cells, effect of infection with iridescent virus on synthesis of 120
— viruses of arthropods 1, 27, 28, 32, 38, 65, 113, 133, 135, 138, 139
dodecahedral inclusion bodies 133
drop size, influence on plant coverage 146
Drosophila spp. 139
— *melanogaster* 59
— —, embryonic tissue culture 108
— —, germ cell formation 61
— —, infection with rabies virus 91
— —, infection with vesicular stomatitis virus 91
— —, influence of genotype to sigma virus multiplication 68
— —, permanent paralysis when exposed to CO_2 59
— —, sigma virus infection of 59
duck embryo cell cultures, use for isolation of arboviruses 51
duration of heat treatment and progressive loss of sigma virus from germ cells 82

Eagles' medium 110
early décrochage 78
eastern encephalitis virus (group A) 43, 45, 46, 49, 51
— equine encephalomyelitis virus, multiplication in primary tick tissue cultures 124
— — — —, multiplication in tissue cultures of *Dermacentor pictus* and *Ixodes ricinus* 124
effectiveness of arthropod viruses as living insecticides 143
Egypt 101 West Nile virus, multiplication in primary cultures of imaginal discs of *Aedes aegypti* larvae 123
electron micrograph of microcrystals of wound tumor virus in cytoplasm of hemocyte of leafhopper 97
— — of pea enation mosaic virus in cytoplasm of fat body cell of *Acyrthosiphon pisum* 103
— — — — — — in gut lumen of the aphid vector *Acyrthosiphon pisum* 103
— — wound tumor virus in the cytoplasm of a hemocyte of *A. constricta* 101
— — — — — particles in cytoplasm of a nerve cell of *A. constricta* 99
— — — — — in tubular structures in the cytoplasm of a nerve cell of *A. constricta* 100
— micrographs of inclusion-type insect viruses 2, 3, 4, 5, 130, 131, 132, 134, 136, 137, 138, 139, 145
— — of noninclusion type insect viruses 131, 140, 141
— — of plant viruses in vector tissues 97, 99, 100, 103, 104
— microscopic observations of inclusion bodies 2
— — study of development of nuclear polyhedrosis virus in tissue culture cells 118
— — — formation of wound tumor virus in insect vectors 102
— microscopy 2, 95, 102, 118, 132
embryonic cells, resistance to nuclear-polyhedrosis viruses 11
— tissue culture of *Aedes aegypti* 108
— — — *Drosophila melanogaster* 108
— — — leafhopper 108
encephalitogenic viruses, transmitted in birds 46
encephalomyocarditis virus 50
environmental persistence 149
— stress factors, influencing the incidence of viral diseases 149
enzymatic treatment of insect tissue 109
epidemiology, human 33
— of non-inclusion type insect viruses 33
epidermis 98, 105, 116

epizootics, importance of adult transmission for 19
—, — parasites for development 20
—, — temperature for development 20
— in *Lepidoptera* 15
—, studies in nuclear-polyhedrosis of *Diprion hercyniae* 18
—, theory of their development 18
Estigmene acrea, tissue cultures of 110
ether soluble lipids in membranes of rod-shaped virions 6
Euproctis chrysorrhoca 151
European foulbrood of the bee 26, 34
evolution, distinct lines in rod-shaped inclusion-type insect viruses 7
experimental paralysis of the bee 29
— transmission of *Tipula* iridescent virus to other insects 142
explants of adult tissue of *Dermacentor marginatus* and *Dermacentor pictus* 110
— of tissue from the cockroach *Blattella germanica* 110
extracts of healthy insects, increasing resistance of *Prodenia litura* against virus infection 151
extrinsic incubation period of arboviruses 45

Fall-beet armyworm complex of viruses 159
fat body 8, 116
— —, one of the likely sites of wound tumor virus multiplication in leafhopper vector 98
— — tissue 96, 102, 103, 105
— — —, changes resulting from wound tumor virus infection 102
fatty acids, increase during polyhedroses 10
Federal Food, Drug and Cosmetic Act of 1938 153
— Insecticide, Fungicide, and Rodenticide Act 153
feeding habits of the target pest as factor of effectiveness of viruses 147
fetal bovine serum, substitute for hemolymph in insect tissue culture medium 110
field use of arthropod viruses 144
fine-structure of inclusion bodies 2
flaccidity of *Antheraea eucalypti* 29
— silkworms 26
— —, pathological findings 26
Flanders virus 43
5-fluorodeoxyuridine, delay of multiplication of sigma virus 65
Food and Drug Administration, tolerances established by the 153
"Forest disease" of the bee 29
formation of wound tumor virus in insect vectors, electron microscopic study 102
foulbrood 25, 32, 34
frequency of CO_2-sensitive progeny after inoculation of different sigma virus genotypes into female flies 76
fungi, interference with inclusion-type insect virus multiplication 17

Galleria mellonella 14, 18, 27, 33, 110, 112, 151
— — cultures of ovarial, fat body and blood cells, production of polyhedra in 112
— —, increased susceptibility to inclusion-type virus by blocking the phagocytes with carbon particles 18
— —, infection with densonucleosis virus 27, 33
— —, temperature dependent increase of the resistance against virus infection 151
— —, tissue cultures of 110
ganglion cells 98
genetic virus, noninfectious sigma virus component confined to the germ line 89
genes causing resistance to sigma virus, localization of 68
genotype of fly, influence on sigma virus multiplication 68
— of sigma virus, dependence of final plateau level of titer on 66
genotypes of sigma virus 76
general virology, role of the invertebrate viruses played in 34
Geometridae 17
germ cell formation in *Drosophila* 61
— — — — related to a model for sigma virus infection 87
— line infection of sigma virus, maintenance of the stabilized state 86
— — invasion by sigma virus, interaction of different virus-genotypes 77

"germinal passage" of sigma virus 76
— — — —, synergistic effects of different virus genomes 81
glial cells 98
gnats as vectors of arboviruses 42, 43
gonadal cell cultures of insects, primary and serially passaged cultures 112
GRACE *Aedes* line, medium for culture of 110
— *Antheraea* line, medium for culture of 110
granules of granulosis virus, fine structure 3
granuloses 1, 134, 135, 138, 140, 141, 144, 152, 158
— as viral insecticides 158
granulosis of the codling moth, *Carpocapsa pommonella* 3
— virus 1, 3, 114, 134, 135, 138, 139, 140, 141, 143, 144, 147, 150, 152, 154, 156, 158
— —, multiplication in insect tissue culture 114
— — of the cabbageworms, industrial production of 156
— — *Carpocapsa pomonella* 4, 9
— — *Choristoneura fumiferana* 4, 9
— — *Pieris brassicae* 9
— — *Pieris rapae* 9
— — *Zeiraphera diniana* 16
— viruses, chemical studies 5
— —, course of infection 9
— —, course of infection related to tissue sites of infection 9
— —, cuboidal variants 4
— —, host response in different virus-host systems 15
— —, industrially produced in living insect culture 158
— —, natural occurence 17, 140, 141
— — of *Lepidoptera* 7
— —, protein of granules 5
— —, release and concentration of virions from inclusion bodies 14
— —, species specificity of the 152
— —, virulence 15
— —, — of viruses from different races of *Pieris brassicae* 15
green monkey kidney cell cultures, use for isolation of arboviruses 51
Gryllus bimaculatus, paralysis of 25
gut 96, 103, 105
heavy water gradient centrifugation 95
HeLa cell cultures, use for isolation of arboviruses 51
Heliothis nuclear-polyhedrosis virus, susceptibility of different species against 151
— *virescens* 16, 147, 150
— —, susceptibility to nuclear-polyhedrosis virus 16
— *zea* 16, 147, 148, 150
hemagglutination-inhibition test used for classification of arboviruses 39
Hemerocampa leucostigma, tissue cultures from 111
hemocyte cytoplasm 97
hemolymph as medium supplement for insect cell lines 109
— supplement for insect tissue culture medium, replacement by calf serum or fetal bovine serum 110
— from *Antheraea pernyi* as medium supplement for cell lines 110, 125
— *Bombyx mori* as medium supplement for insect tissue cultures 110
— of different insect species, unspecificity for medium supplement in insect tissue culture 110
— insects, differences in composition to mammalian blood 110
— —, level of amino acids 110
— —, level of calcium and magnesium ions 110
— —, pH 110
— —, presence of organic phosphates, organic acids and trehalose in 110
— —, sodium-potassium ratio 110
— of leafhopper vectors, carrier and medium for multiplication of virus 102
hereditary transmission of sigma virus 76
— — — —, diagrammatical model 88
— — — —, influence of virus genome 76
— — — —, relationship to content of infectious material in flies 84
hibernating rodents as arbovirus reservoirs 48
horizontal transmission of inclusion-type insect viruses 8

host plant, influence on the incidence of polyhedroses 148
— range of non-inclusion type viruses of invertebrates 30
— response to granulosis viruses in different virus-host systems 15
— specificity of insect viruses 13, 115
— — — — in vitro 115
— spectra of arboviruses 46
human health, role of the U.S. Department of Interior review on the registration-application for 153
humidity, influence on field applied viruses 150
humoral factors in limitation of sigma virus multiplication in stabilized flies 89
Hyalomma asiaticum, multiplication of TE virus, strain Hypr in tissue cultures from 124
— *dromedarii* tissue culture, multiplication or survival of arboviruses group *A* and *B* and non-arboviruses in 125
hyaluronidase treatment 109
hydrolytic enzymes from hepatopancreas and crop of *Helix aspersa*, use for insect tissue culture 109
hygiene of useful invertebrates 34
Hymenoptera 1, 16, 25, 26, 29, 32, 140, 151
—, inclusion-type viruses 16, 140
—, non-inclusion type viruses 25, 26, 29, 32, 140
Hyperomyzus lactucae, systemic infection by lettuce necrotic yellows virus 105
Hyphantria cunea 151
hypotheses on sigma virus infection of *Drosophila melanogaster* 84

icosahedral inclusion bodies 133
— viruses 25, 26, 27, 28
imaginal buds 116
— fecundity, lowering by sigma virus infection 62
inactivation of nuclear-polyhedrosis virus in midgut 14
inapparent infection with densonucleosis, serological detection 27
inclusion bodies, behavior in weak alkaline 4, 6
— —, characterization 1
— —, chemical studies 4
inclusion bodies of insect viruses, preparation of pure suspensions 13
— — — —, size and shape in insect tissue cultures 119
— —, resistance to putrefaction 13
— —, size 1, 135
— —, type 1, 135
"inclusion body" 35
— — insect viruses, growth in insect tissue culture 111, 112, 113
— — — —, multiplication of infectious particles and typical polyhedra formation in primary ovarian tissue culture of *Bombyx mori* 111
— — viruses 1, 111, 113, 130, 132, 135, 138, 140, 144, 158
inclusion-type insect viruses 1, 134, 135
— — — —, chemistry of 1
— — — —, concentration of free virions by ultracentrifugation 14
— — — —, detection of disease in pupal and adult stage 10
— — — —, development in the infected host 8
— — — —, different types 1, 134, 135
— — — —, different ways of infection of host insects 8
— — — —, DNA nucleoprotein containing 1
— — — —, incidence of disease in the larval, pupal and adult stages 10
— — — —, interactions of different virus types 17
— — — —, — with bacteria, protozoa and fungi 17
— — — —, isolation from species of the orders *Lepidoptera, Hymenoptera, Diptera, Orthoptera* and *Neuroptera* 16, 140
— — — —, multiplication in insect tissue culture 111
— — — —, natural occurence and dispersal 16, 140
— — — —, nucleic acids of 6, 135
— — — —, preparation of pure suspensions of inclusion bodies 13, 113
— — — —, presence in cells in an occult state 12, 114
— — — —, process of infection 8
— — — —, quantitative virus-host-relationships 13

inclusion-type insect viruses, release of infectious virus in the environment 8, 9, 10
— — — —, — of virions from inclusion bodies by alkali treatment 13, 112
— — — —, RNA nucleoprotein containing 2
— — — —, route of contamination of the environment by infected hosts 8, 9
— — — —, size 6, 135
— — — —, structure of 1, 135
— — — —, survival in hosts in an inapparent state 12
— — — —, survival in hosts without pathologic or cytopathic changes 12
— — — —, survival outside their hosts 13
— — — —, susceptible tissues to 8, 9, 116
— —viruses, differentiation of 133, 135
inclusions in cells infected by insect viruses 1, 112, 130, 135
incubation time, a characteristic for defining sigma virus strains 70
— — of sigma virus infection 63
— — — — —, relation to number of infectious units injected 64
industrial propagation of bollworm nuclear polyhedrosis 156
— — codling moth granulosis 156
— — pine processionary moth cytoplasmic polyhedrosis 156
— — of *Tipula* iridescent noninclusion virus 156
infection process of granulosis virus, quantitative study 14
— test of cabbage looper nuclear polyhedrosis virus on guinea pigs and mice 155
— — cotton bollworm nuclear polyhedrosis virus on guinea pigs and mice 155
— — *Heliothis* virus by feeding to men 155
— threshold phenomenon 44
infectivity in insect tissue cultures of purified virus rods mixed with polyhedron protein 113
— — — — virus rods separated from polyhedron protein 112
— of nuclear polyhedrosis viruses during hosts diapause 11
infectivity titers of hemolymph from polyhedrosis diseased insects, comparison in insect and insect tissue culture 113
— — virus rods from polyhedrosis viruses in tissue culture and in insects 113
insect-associated viruses, in vitro studies 111
— — —, use of insect tissue culture for study of 108
— blood cells, ability to divide in vitro 109
— — —, conditions for cultivation 109
— — —, content of polyphenoloxidase 109
— — —, intracellular melanization 109
— — —, production of toxic quinone substances in presence of oxygen in culture by 109
— — —, susceptibility to insect viruses 109
— cell cultures, primary 108
— — line from the mosquito *Aedes aegypti* 109
— — — moth *Antheraea eucalypti* 109
— — lines 109
— — —, need of hemolymph as medium supplement 109
— — —, subline without need of hemolymph as medium supplement 109
— cells, damage by trypsin of 109
— embryonic tissue culture 108
— — — —, methods for preparation 108
— hemolymph, differences in composition to mammalian blood 110
— —, level of amino acids, calcium and magnesium ions in 110
— —, pH 110
— —, presence of organic phosphates, organic acids and trehalose in 110
— —, sodium-potassium ratio 110
— ovarian and epidermal tissue, dispersion of 109
— -pathogenic viruses, in vitro studies 111
— tissue culture medium, hemolymph, chick embryo extract, calf or fetal bovine serum as supplement 110
— — cultures 108

insect tissue cultures, contribution of polyhedron protein to infectivity of virus rods 113
— — —, cultivation of arboviruses in 122
— — —, cultivation of plant viruses in 121
— — —, electron microscopic study of development of nuclear polyhedrosis virus in 118
— — —, explants 110
— — —, handicaps 108
— — —, infection with cytoplasmic polyhedrosis virus from *Porthetria dispar* 114
— — —, — hemolymph of diseased insects 111
— — —, — inclusion body viruses 111
— — —, — purified noninclusion insect viruses 114, 115
— — —, — purified virus material from alkali dissolved polyhedra 112
— — —, LD_{50} of dissolved washed polyhedra compared with the LD_{50} in insects 113
— — —, — polyhedrosis viruses in 113
— — —, medium requirements 110
— — —, multiplication and passage of polyhedral virus in 111
— — —, multiplication of arthropod viruses in 111, 158
— — —, pathological signs of infection with nuclear polyhedrosis virus 118
— — —, phases of development of insect viruses in 117
— — —, problems involved in 108
— — —, serum supplements of medium for 110
— — —, size and shape of polyhedra in 119
— — —, use for isolation of arboviruses 51
— — —, — study of insect-associated viruses 108
— — —, — — plant virus-vector relationship 121
— — —, value for investigations in insect viruses 120
— tissues, dissociation with hydrolytic enzymes from hepatopancreas and crop of *Helix aspersa* 109
— —, enzymatic treatment 109
— —, treatment for establishment of primary cultures 109
insect virus study group of the virus subcommittee on bacteriological nomenclature 142
— viruses 1, 24, 59, 94, 108, 129
— —, inclusion-type 1, 111, 129, 134, 135, 140, 158
— —, non-inclusion type 24, 114, 130, 134, 135, 140, 158
— —, phases of development in vitro 117
insecticides, use of viruses as 129
insects as intermediate hosts and vectors for viruses 108
— hosts and vectors for arboviruses, criteria for establishment of this relationship 41, 122
— vectors, alternate hosts and reservoirs of plant viruses 94
—, diverse relationships of viruses to 108
— infected by plant viruses 94
"integrated" sigma virus 88
interactions of bacteria, protozoa and fungi with inclusion-type insect viruses 17
interference at the cellular level of granulosis and nuclear-polyhedrosis virus 17
invertebrates 1, 24, 40, 59, 94, 108
in vitro infection of *B. mori* ovarian tissue culture with nuclear polyhedrosis virus 112
— — studies of insect-associated viruses 111
iridescent viroses 27, 33, 114, 142, 154, 156, 158
— —, pathological findings in 27, 28
— virosis, properties of the virus 28
— —, serological differentiation of similar viruses 28
— —, spectrum of virulence 28
— viruses, detection in insect tissue culture by fluorescent antibody staining 114
— —, development in primary cultures of gonads, circulatory organs, fat body and brain 116
— —, multiplication in insect tissue cultures 114
— —, phases of development in insect tissue cultures 119
— —, sites of development in vivo 28, 116, 134

isolation of arboviruses by intracerebral infection of suckling mice 50
Ixodes ricinus, multiplication of EEE virus in cultures of fat body and hypodermis tissue of 124

Japanese B — West Nile complex of arbovirus group B 39, 41, 46, 49
— encephalitis virus, multiplication in Grace *Antheraea* cell line 125
Junin virus 43, 49, 50

Kyasanur Forest tick fever virus 49, 52
kind and number of arthropod viruses 140

"laboratory virosis" of the bee 29
lactalbumin hydrolysate in medium for insect tissue cultures 110
larch budmoth, *Zeiraphera diniana*, granulosis virus infection in 14
larval maturation and development of resistance against viruses 150
— stage, differences in susceptibility to polyhedroses 14
Lasiocampidae 17
late décrochage 78
latent infection of polyhedra in tissue cultures from *Hemerocampa leucostigma* 111
— virus infection 90
lateral nerves 98
L cell cultures, use for isolation of arboviruses 51
LCM virus, evidence for multiplication in *Hyalomma dromedarii* tissue culture 125
LD_{50} of dissolved washed polyhedra in insects and insect tissue cultures 113
— polyhedrosis virus in hemolymph of diseased insects, comparison in insect tissue culture and insects 113
— sacbrood virus of the bee 25
— units of nuclear-polyhedrosis viruses in different hosts 15, 143
leafhopper, primary culture of embryonic tissue 108
— -borne viruses, changes in virus-transmitting individuals 96
— — —, congenital acquirement 96
— — —, lethality to embryos, nymphs or adult vectors 96
— — —, resemblance and differences to sigma virus of *Drosophila* 96
leafhopper-borne wound tumor virus, electron microscopic study 96
leafhoppers as vectors for plant viruses 94
Lepidoptera 1, 28, 32, 109, 140, 151, 152
—, inclusion-type viruses 16, 140
—, increase of virus resistance with increasing larval age 14
—, non-inclusion type viruses 28, 32, 140
—, primary tissue cultures of 109, 110
lethargy of *Melolontha melolontha* 28
— — —, pathological findings 28
lettuce necrotic yellows virus, systemic infection of *Hyperomyzus lactucae* aphids 105
liquefaction of host and release and survival of nuclear-polyhedrosis and granulosis viruses 12
living insects as source of viruses 158
localization of densonucleosis 27
— development of inclusion-type insect viruses in the infected host 8
— non-inclusion type viruses of invertebrates in the cell 30
long term control of *Diprion hercyniae* by polyhedrosis virus 150
— — cultivation of blood cells of rice stem borer *Chilo suppressalis* 109
loss of arthropod vector by U.S. bat salivary gland virus 50
louping ill virus 39
lumen of the midgut, role in resistance against polyhedroses 14
Lymantridae 17

Machupo virus 43, 49, 50, 54
Macropipus depurator, paralysis of 26
Macrosteles fascifrons, multiplication of aster yellows virus in 95
magnesium ions, level in hemolymph 110
maintenance of mixed infection in rho-infected flies 79
Malacosoma, virulence of nuclear-polyhedrosis viruses to different species of 15
— *americanum*, tissue cultures of 110
— *disstria*, infectibility with nuclear polyhedrosis virus from other species 150
— —, susceptibility of larvae to polyhedra 14

Malaya disease of *Oryctes* 31, 33
Malpighian tubules 96
mechanical inoculation as assay of plant virus multiplication in insects 95
mechanisms and control of formation of inclusion bodies 3
media for vertebrate tissue cultures, use for insect tissue cultures 110
medium and supplements for insect tissue culture 110
— for explant cultures from insect tissues 110
— growth of insect cell lines 109, 110
— — of insect tissue cultures 110
— requirements for tissue cultures of phytophagous, zoophagous and omnivorous insects 110
melanization, intracellular of blood cells of insects 109
Melolontha melolontha 25, 138
— —, lethargy of 28
— —, transparent disease of 33
methods for preparing primary cultures of insect embryonic tissue 108
— of studying multiplication of plant viruses in insect vectors 95
microcrystals of wound tumor virus in cells and tissues of leafhopper vector 96
mites, probably vectors of arboviruses 42, 43
Miller Pesticide Chemical Amendment of 1954 153
models for sigma virus infection 84, 86, 87
mole ratio of sodium to potassium, range in hemolymph of insects 110
Morator viruses 25, 134, 142
Moratorvirus lamellicornium 25
MORISON's inclusions 29
morphology of non-inclusion viruses, compared with plant and animal viruses 132, 133, 139
— the virions of inclusion-type insect viruses 6, 135
— — non-inclusion type viruses 32, 135
mortality in sigma virus infected *Drosophila melanogaster* 62
mosquitoes as vectors of arboviruses 42
—, infected with iridescent virosis 28
multihit curves of heat treatment of sigma virus infection 82
multiplication curves of sigma virus 66
— of granulosis virus in susceptible tissue 14
— plant pathogenic viruses in aphids 103, 105
— — — — insect vectors 94
— — — — —, assay methods 95
— plaque size types of sigma virus in flies 71
— sigma virus 66
— — —, growth curves 66
— — —, influence of fly and virus mutations 68
— — —, influence of temperature 72
multiplicity of SIV infection 115
Musca domestica 151
muscle cells 96, 105
mutation rates of sigma virus 69, 79, 80
mutations of fly lending resistance to sigma virus infection 68
mycetome 98, 105
Myzus persicae, changes in stomach cells after infection with plant viruses 102

natural groups of arthropod viruses 142
— transmission of arthropod viruses 152
Nauphoeta cinerea 151
Neodiprion 147, 148, 150
— *swainei* 152
Nephotettix cincticeps embryonic tissue culture, replication of rice dwarf virus in 121
— — vector of rice dwarf virus 98
nerve cells 97, 98, 116
nervous system of *Agallia constricta*, infection with wound tumor virus 98
Neuroptera 16, 140, 142
—, inclusion-type viruses 16, 140
neutralization test, used for classification of arboviruses 39
Noctuidae 17
nomenclature of arthropod viruses 142
Nomia melanderi 139
Nomophila noctinella 151
non-arboviruses tested for multiplication in *Hyalomma dromedarii* tissue culture 125
non-inclusion body viruses, phases of development in insect tissue cultures 119

non-inclusion insect viruses, development in insect tissue cultures 114, 158
— nuclear insect viruses, lack of host specificity in vitro 116
— type insect viruses, structure of 25, 26, 27, 32, 135
— viroses, industrially produced in living insect cultures 158
— —, industrially produced in tissue-cell culture 158
— —, lack of data for use as insecticides 159
— viruses 24, 32, 114, 130, 132, 134, 135, 139, 140, 141, 142, 143, 144, 151, 154, 156, 158
non-stabilized state of sigma virus infection 85
non-susceptibility of embryonic cells to nuclear polyhedrosis viruses 11
nuclear inclusion bodies 1, 111, 130, 134, 135
— noninclusion virus, development in primary insect tissue cultures 115
— polyhedra, amino acid content of the protein 5
— —, chemical studies of 4
— —, molecular weight of the protein 5
— — of *Lepidoptera* 7
— polyhedroses 1, 116, 133
— — as viral insecticides 143, 158, 159
— polyhedrosis of the greater wax moth, *Galleria mellonella* 2, 112
— —, rearing of disease-free populations 19
— — virus, completion of development in vitro in explants from diseased silkworm 112
— — —, cytopathic changes in insect tissue cultures 112
— — — of alfalfa caterpillar, industrial production of 156
— — — *Bombyx mori*, infections for *Galleria mellonella* 15
— — — — —, selection of a strain producing cuboidal inclusion bodies 119
— — — cabbage armyworm, industrial production of 156
— — — cabbage looper, industrial production of 156
— — — cotton bollworm, industrial production of 156
nuclear polyhedrosis virus of *Diprion hercyniae* 15, 144
— — — European spruce sawfly, *Diprion hercyniae*, study of the virus-host system 18
— — — Great Basin tent caterpillar, industrial production of 156
— — — *Heliothis* 16, 144, 156
— — — *Malacosoma disstria*, experimental field studies 19
— — — *Peridroma saucia*, multiplication in blood cell cultures 112
— — — sawflies, industrial production of 156
— — — silkworm, infection of *B. mori* ovarian tissue culture in vitro 112
— — — *Trichoplusia ni* 15, 144
— — —, response in geographically separated populations of California oakworm, *Phryganidia californica* to 16
— — viruses 1, 15, 16, 111, 112, 116, 117, 130, 134, 135, 141, 145, 156, 158
— — —, adsorption to cells in suspension 117
— — —, capability of infecting closely-related host species 15
— — —, course of infection in lepidopterous larvae 8
— — —, course of infection in *Hymenoptera* 9
— — —, development in different hosts 15
— — —, difference of activity peaks in exchange column of hemolymph and dissolved polyhedra preparations 113
— — —, egg-transmission of 19
— — —, electron microscopic study of development in tissue culture of 118
— — —, histopathological changes in adults of both sexes 19
— — —, importance of adult transmission for epizootics of 19
— — —, — parasites for development of epizootics 20
— — —, — temperature for development of epizootics 20
— — —, infection in adults of both sexes 19
— — —, lack of host specificity in vitro 116

nuclear polyhedrosis viruses, means of dispersion 19
— — —, multiplication in blood cell cultures 116
— — —, — cultures of fat body cells 116
— — —, — — ovarian tissue 116
— — —, natural occurence 17, 140, 141, 144
— — — of *Hymenoptera* 7, 9, 16, 140, 144
— — — *Lepidoptera* 7, 19, 140, 144
— — — —, a closely-related group 7
— — — sawflies 18, 19, 141, 144, 156
— — —, pathological signs of infection in tissue culture cells 118
— — —, purification from hemolymph with Sephadex cation exchange column 113
— — —, release and concentration of virions from inclusion bodies 13
— — —, resistance of embryonic cells to 18
— — —, — larvae at different temperature 20
— — —, rod shape and DNA content 113
— — —, sites of multiplication in vivo 8, 116
— — —, survival and spread by mechanical contamination of foliage 19
— — —, — in the host population 19
— — —, susceptibility of different larval stages to 19
— — —, transmission 19
— — —, transmission by both sexes 19
— — —, virulence of 18
nuclei of plant cells infected with pea enation mosaic virus, accumulations of virus particles in 105
— stomach cells, enlargement in virus infected aphids 102, 105
nucleic acid of sacbrood virus 25
— acids of inclusion-type insect viruses 6, 135
— — noninclusion-type insect viruses 25, 32, 135
Nymphalidae 17

"occult" viruses 152
Olethreutidae 4, 17
omnivorous insects, medium requirements for tissue cultures of 110
Omsk tick fever virus 49
O'nyong-nyong fever 52
oocyte invasion by sigma virus 61
— — — —, influence on fecundity and fertility of females 77
oogenesis in *Drosophila* 62
oogonia, infected by sigma virus 61
oogonial infection with sigma virus, content of infectious material in flies with 84
— — — —, immunity to superinfection 61
— — — —, latent state 61
— — — —, ultra-rho condition 61
Opherophtera brumata 138
organic acids, presence in hemolymph 110
— phosphates, presence in hemolymph 110
origin of arboviruses 49
Orthoptera 16, 32, 94, 140, 142
—, inclusion-type viruses 16, 140
—, viral disease in 25, 32, 95, 140
Oryctes rhinoceros, Malaya disease of 31, 33
Ostrinia nubilalis 151
ovarian cells of insects primary and serially passaged cultures, use for infection with nuclear polyhedrosis virus 112
ovaries 108, 116
ovary culture of insects 108
— of *Bombyx mori* pupae, tissue culture of 108

Paillotella viruses 134, 142
Panonychus citri 27, 33, 146
— —, control by means of free virus of 33
— *ulmi* 27
Paralysis — black disease of the bee 29
— — — — —, histopathology 29
— in *D. melanogaster* infected with sigma virus, specifically induced by CO_2 59
— of bee 34
— of crabs 26
— —, pathological findings in 26
— —, spectrum of virulence 26
— crickets 25
— —, pathological findings 25
parasites, importance in dispersal of nuclear-polyhedrosis virus 20
paraspherical viruses 25, 26, 35

parenteral infection of insects with arboviruses 49
particle counting method for assay of plant virus multiplication in vectors 95
pathogenesis of non-inclusion type viruses of invertebrates 30
pathogenicity of arboviruses for suckling mice 39, 42
— — in vertebrates 49
— of viral insecticides 153
— of Western-X-disease virus to the leafhopper 96
pathological findings in arthropods infected with polyhedroses 8, 133
— — in densonucleosis 27
pea aphids, localization of pea enation mosaic virus in 103
— enation mosaic virus 102
— — — —, accumulation of virus particles in nuclei of infected plant cells 105
— — — —, electron micrographs 103, 104
— — — —, — microscopic localization in infected pea plants and pea aphids 103
— — — —, presence in nuclei 105
— — — —, sequential invasion of various cell structures by 105
— plants, electron microscopic localization of pea enation mosaic virus in 103
Pectinophora gossypiella, cytoplasmic polyhedrosis virus, transmission experiments 151
Peridroma margaritosa, influence of environmental factors on the granulosis of 149
— *saucia* blood cell cultures, infection with the nuclear polyhedrosis virus 112
— —, conditions for cultivation of blood cells of 109
— —, culture of thoracic tissue 109
perineurium cells 98
Periplaneta americana 151
persistence of cytoplasmic polyhedrosis virus of pine processionary moth on treated plants 150
— granulosis virus of cabbage worm on treated cabbage plants 150
— sigma virus in non-infectious form during heat treatment 75
persistence of viral insecticides in the blood of vertebrate test animals 154
— — — feces of vertebrate test animals 154
— — — tissue of vertebrate test animals 154
— virus as latent infection in overwintering stages of arthropods 149
— — in living adult arthropods 149
persistent circulative viruses 102
pH, influence on virions and inclusion bodies of 148
— of hemolymph 110
Phalaenidae 17
phase changes of sigma virus characters 81
phenotypes of sigma virus characterized by differential multiplications rates 70
Phlebotomus 43
— group of viruses 43
phospholipids, increase during polyhedroses 10
Phryganida californica 150
physiological changes in host larvae during development of nuclear and cytoplasmic polyhedroses 10
— — — — infected with nuclear and cytoplasmic polyhedroses 10
physiological stress, influence on the incidence of polyhedroses of 148
phytophagous insects, medium requirements for tissue cultures 110
Pieridae 17, 147
Pieris brassicae 147
— *rapae* 138, 147, 148
pine processionary moth cytoplasmic polyhedrosis 156
— — — — —, safety to various organisms 154
plant cell nuclei, accumulation of pea enation mosaic virus particles in 105
— pathogenic viruses, host range 94
— — — in insects 94
— — —, interaction with aphid vectors 102
— — —, multiplication in aphids 103, 105
— — —, systemic infection of an aphid by 104
— virus diseases 94
— viruses, congenital infection of leafhopper progeny with 94

plant viruses, cultivation in insect tissue cultures 121
— —, development in cultures of vector tissues 121
— —, use of insect tissue culture for study of virus-vector relationship 121
"plaque size" of sigma virus strains 70
— — types of sigma virus, multiplication in flies 71, 79
— type of sigma virus, measurement of 70
plasmagenes as "occult" viruses 153
Plusia gamma 150
polyhedra, content of silicon 6
— formation in infected primary cultures of ovarian tissue from *Bombyx mori* 111
— — tissue cultures from *Hemerocampa leucostigma* 111
—, genetical control of shape by the virus 119
— latent infection in tissue culture, manifestation after medium change 111
— production in tissue culture of ovarial, fat body and blood cells of *Galleria mellonella* 112
—, shape in tissue culture cells and in insects 1, 119
—, size in tissue culture cells and in insects 1, 119
polyhedral inclusion bodies 1, 24, 111, 119, 130, 133
— — —, dimensions of 1, 119, 133
— protein, molecular weight and composition 5
— —, site of formation 10
polyhedron, fine-structure 2
— protein, contribution to infectivity of virus rods in insect tissue cultures 113
polyhedroses 1, 111, 133, 134, 135, 144, 154, 156, 158
— of silkworm 8, 11, 15, 24, 111
polyhedrosis viruses 1, 111, 130, 133, 134, 135, 136, 140, 141, 143, 145, 147, 150, 151, 152, 154, 156, 158
— —, industrially produced 158
— —, infection of insect tissue cultures with purified virus rods 112
— —, infectivity of purified virus rods mixed with polyhedron protein in insect tissue cultures 113
polyhedrosis viruses, LD_{50} in insect tissue culture and insects 14, 113, 143
— —, — of dissolved washed polyhedra in insects and insect tissue cultures 113
— —, virus rods separated from polyhedron protein, infectivity for insect tissue culture 112
polymorphic inclusion virus 134, 135, 138, 140, 141, 142
polyphenol-oxidase, content in insect blood cells 109
population density, influence on the incidence of polyhedroses 148
— response to granulosis virus 16
— — nuclear-polyhedrosis viruses 16
porcine kidney cell cultures, used for isolation of arboviruses 51
Porthetria dispar 114, 150
— —, cytoplasmic polyhedrosis virus infection 114
— — primary cell cultures, infection with nuclear polyhedrosis virus from *B. mori* and *A. pernyi* 116
potato leafroll virus 102
preparation of insect viruses for field use 149
prepupae of *Diprion hercyniae*, not susceptible to viral infections 151
primary cultures of insect embryonic tissue, preparation 108
— — leafhopper embryonic tissue 108
— — tissue from phytophagous, zoophagous and omnivorous insects, medium requirements 110
— embryonic cell culture of *Aedes aegypti* 108, 110
— hamster kidney cell cultures used for isolation of arboviruses 51
— insect cell cultures 108
— tissue cultures of *Lepidoptera*, medium requirements 110
Prodenia eridania 151
— *litura* 152
— —, increase of resistance against infection by means of extracts of healthy insects 151
production and industrial feasibility of viral insecticides 155
propagation of various types of arthropod viruses 156
properties of the virus of iridescent virosis 28

protein of granulosis virus granules, molecular weight 5
— of polyhedra, molecular weight and composition 5
protozoa, interactions with inclusion type insect viruses 17
Pseudaletia unipunctata, synergistic reaction of granulosis and nuclear-polyhedrosis viruses in 17
Pseudomorator virus 134
Psychidae 17
Public health importance of arboviruses 51
purification of inclusion bodies 13, 113
— of sigma virus 65

quantitative studies on infection process in inclusion-type insect viruses 14, 117

rabies virus, multiplication in *Drosophila melanogaster* 91
reactivation of heat inactivated sigma virus 73, 74
recombinants of sigma virus, arise of segregants 79
— — —, clonal distribution of 79
— — —, stability 79
recombination, frequency after mixed infection of different strains of sigma virus 78, 79
RDV 121
"refractory" gene, influence on sigma virus multiplication 68
re genotype of *Drosophila melanogaster* 69
registration of field applied viruses with United States Department of Agriculture 153
release of infectious virions from polyhedra in vivo 6
— virions from inclusion bodies by alkali treatment 13, 112
reptiles as hosts of arboviruses 46, 47, 48
reservoirs of arboviruses in vertebrates 48
residues, remaining on food or live stock products after application of viral insecticides 153
resistance development against viruses by chemicals 151
— — — by temperature 151
resistance factors to polyhedroses, increase with larval age 14
— of arthropods to virus 150
— *Drosophila* to sigma virus multiplication, different causes 84
— embryonic regenerative cells to nuclear polyhedrosis viruses 11
— to cytoplasmic polyhedrosis virus in silkworm, control by a dominant majorgen 15
retardation of development in sigma virus infected *Drosophila melanogaster* 62
rho behavior of sigma virus 69
— infection 71
— —, effect of temperature on the transmission of 83
— —, influence of temperature on 75
— variant, limits of hereditary transmission by the mutant fly gene re 76
— — of sigma virus 71
— — — —, CO_2-sensitivity in infection 71
rice dwarf virus, changes in infected cultures of embryonic leafhopper cells 121
— — — development in embryonic tissues from eggs of viruliferous femals, electron microscopic study 121
— — —, multiplication in its insect carrier 95
— — —, replication in embryonic tissue culture of rice leafhopper 121
— — —, study of transovarial transmission in embryonic tissue cultures from eggs of viruliferous females 121
— leafhopper, transovarial serial passage of rice dwarf virus in 95
RNA content of spherical or subspherical membraneless virions of cytoplasmic polyhedrosis 6
—, increase in midgut cells during cytoplasmic-polyhedroses 10
— -non-inclusion viruses of invertebrates 25, 32, 135
— viruses of arthropods 2, 25, 32, 38, 42, 65, 113, 133, 135, 139
rodents as hosts of arboviruses 46, 49, 50
— — Colorado tick fever virus 49
— — Venezuelan encephalitis virus 49
rod-shaped virions 1

rough dwarf virus of maize 102
route of infection of inclusion-type insect viruses 8

sacbrood of the bee 25, 34
— —, pathological findings 25
safety of field applied viruses to animals and plants 153
— several types of arthropod viruses to plants, beneficial arthropods, vertebrates and humans 154
— regulations for viral insecticides 153
— test of viral insecticides by feeding to vertebrate animals 153
— — — — inhalation to vertebrate animals 153
— — — — injection into various organs of vertebrate animals 153
salivary gland 96, 105
— —, accumulation of rice dwarf virus in 98
— — changes in *Aedes aegypti*, infected with Semliki Forest virus 45
— — — *Culicoides*, infected with blue tongue virus 45
— — viruses of U.S. bat 39
sandflies as vectors of arboviruses 42, 43
Sarcophaga aldrichi, importance in development of epizootics 20
sawflies 1, 11, 15, 141
—, susceptibility to nuclear polyhedrosis viruses 15, 141, 144
— of the genus *Diprion*, susceptibility to nuclear-polyhedrosis viruses 16, 144, 147
— — *Neodiprion*, susceptibility to nuclear-polyhedrosis viruses 16, 144, 147
sawfly complex of viruses 141, 159
Scarabeidae, transparency of 25
Semliki Forest virus 39, 45, 125
sensitivity of the in vitro assay system for SIV compared with the in vivo system 115
Sephadex cation-exchange column, difference between polyhedrosis virus preparations from hemolymph or from dissolved polyhedra on 113
sequential invasion of various cell structures by pea enation mosaic virus 105
— stages of wound tumor virus development in fat body tissues of insect vectors 102
Sericesthis iridescent virus (SIV), asynchronous initiation of viral replication in tissue cultures 120
— — —, check of multiplicity by fluorescent antibody staining 114
— — —, eclipse phase 119
— — —, effect of infection on synthesis of cell DNA 120
— — —, — temperature on 115
— — —, growth in *Antheraea* cell line 114
— — —, yield of infectious units per cell 119
— *pruinosa* 28
serologic similarity of inclusion protein in different host species 7
serological cross-reactions of arboviruses 39
— detection of inapparent infection with densonucleosis 27
— diagnosis of clear head disease of silkworms 26
— differences between viral nucleoprotein and inclusion protein 7
— — of virions from different host species 7
— differentiation of the viruses of iridescent virosis of *Tipula paludosa* and *Sericesthis pruinosa* 28
— identification of different strains of the sacbrood virus of the bee 25
— relationships between inclusion-body protein and virions from different host species 6
— — nuclear-polyhedrosis and granulosis viruses 6
— — of members of arbovirus group B 39
serum supplements for insect tissue culture medium 110
shape of arboviruses 38
— and size of inclusion type insect viruses 1, 135
— — non-inclusion type viruses of invertebrates 25, 26, 27, 35, 135
sigma virus, assay 63
— —, association of infectivity and cell particulates 65
— —, — with a cell organelle 88
— —, attachement to chromosomes 88
— —, centrifugation in sucrose and CeCl gradients 65
— —, cloning technique by germinal passage 81

sigma virus, description of the virion 64
— —, diagrammatical model of hereditary transmission 88
— —, different techniques to get mixed infections 78, 79
— —, "dimensional type" of the strains 70
— —, duration of incubation time as characteristic for different strains 70
— —, electron micrographs 64
— —, experiments with mixed infection of strains carrying different markers 78
— —, frequency of recombination after mixed infection 78, 79
— —, genotype g^+ 76
— —, — g^+ pr^+, frequency of CO_2-sensitive progeny 77
— —, — g^+ pr^-, frequency of CO_2-sensitive progeny 76
— —, — g^- 76
— —, germ line infection 87, 89
— —, heat inactivation 73
— —, hereditary transmission, influence of virus genome 76
— —, — —, limits by the mutant fly gene re 76
— —, — — of mixed infections 79
— —, incomplete or immature particles 90
— — infection, arise of non-stabilized flies from stabilized females 85
— — —, dependence of stabilized and non-stabilized state on source of infection 86
— — —, need of studies on the cellular level 90
— — —, non-stabilized state 85
— — —, occurrence of neostabilization from non-stabilized state 85
— — — of *Drosophila melanogaster* 59
— — — — —, a hereditary infection 59
— — — — —, CO_2-sensitivity 59
— — — — —, CO_2-sensitivity of the imago 63
— — — — —, discovery 59
— — — — —, hereditary transmission 61, 76
— — — — —, history of research on 59
sigma virus infection of *Drosophila*, hypotheses concerning the infection 83
— — — — —, incubation time and infectious units injected 64
— — — — —, interactions between virus and host genoms 59
— — — — —, lowering of imaginal fecundity 62
— — — — —, mortality and retardation of development 62
— — — — —, preview of infection 61
— — — — —, prospects for further investigations 90
— — — — —, response of infected flies 61
— — — — —, transmission by both egg and sperm 59
— — — — —, unusual aspects of the infection 83, 84
— — — — —, virus multiplication 61, 66
— —, infection of flies by inoculation of infectious extracts 61
— —, — oocytes 77
— —, — zygotes by virus carried in sperm 61
— — —, stabilized state 85, 86, 87
— —, infectivity associated with a particulate fraction in homogenates 81
— —, invasion in developing oocytes 61
— —, lability of some characters in different strains 69
— —, latent infection 61
— —, limits of hereditary transmission by the mutant fly gene re 76
— —, maintenance of mixed infection at an equilibrium 79, 80
— —, measure of growth rate of virus strains 70
— —, measure of plaque type 70
— —, methods of concentration 65
— —, — purification 65
— —, models for the infection 84, 86, 87
— —, morphology and dimension of the virion 64
— — multiplication, differences in growth curves in infection via sperm and via egg 67
— — —, growth curves 66
— — — in cell culture 91

sigma virus multiplication in imaginal discs implanted in imagos 91
— — — offspring from crosses of females with oogonial infection and virus-free males 66
— — — — — virus-free females and transmitting males 67
— — —, influence of fly and virus mutations 68
— — —, — fly genotype on 68
— — —, — temperature 72
— —, — of plaque size types in flies 71
— —, mutation rates 69, 79, 80
— — mutations, influence on virus multiplication 70
— —, new types derived from recombination in mixed infections 79
— —, noninfectious component responsible for hereditary transmission of stabilized state 89
— —, nucleic acid content 65
— —, number of virus particles per germ cell 80, 83
— —, oogonial infection 61
— —, — — and production of infectious material in flies 84
— —, partial diploidy of each virus particle 79, 80
— —, persistence in non-infectious form during heat treatment 75
— —, phenotypes with differential multiplication rates 70
— —, phenotypic differences among virus strains 69
— —, physical characteristics of the virion 64
— —, "plaque size" of the strains 70
— —, P^+ strains 70
— —, P^- strains 70
— —, purification by differential centrifugation procedure 65
— —, reactivation after heat treatment 73, 74
— —, relationship of hereditary transmission to content of infectious material in flies 84
— —, rho behavior 69
— —, rho variant 71
— —, stability of different strains 69
— —, stabilized state in germ and somatic cells 86, 87
— —, "strong" strains 70, 82
— —, studies on strains carrying different markers 78
sigma virus, temperature resistant (Tr) strains 70, 73
— —, — sensitive (Ts) strains 70, 72
— —, titration 64
— — transmission, influence of fly and virus mutations on 76
— — —, — of temperature 81
— —, — of infections mixed for plaque size type 79, 80
— —, — virus genomes as groups 79, 81
— —, type v^+ 76
— —, — v^- 76
— —, — — exclusion from sperm 76
— —, — of virus in progeny after mixed infection 78, 79
— —, ultra-rho behavior 69
— —, ultra-rho variant 72
— —, variability of phenotype during multiplication in somatic cells 81
— —, "weak" strains 70, 82
silicon content in polyhedra 6
silk glands 116
silkworm 8, 15, 26, 27, 108, 111, 112, 119, 152
similarity in heat sensitivity of transmission of sigma, rho, and ultra-rho infection 83
— of the virions of nuclear-polyhedrosis and granulosis viruses 6
simultaneous occurence of different inclusion-type insect viruses in the same host 16
sites of virus multiplication in insect cells 95
SIV 28, 32, 114, 115, 119, 120, 158
size of arboviruses 38, 42
— inclusion bodies 1, 135
— inclusion-type insect viruses 6, 135
— non-inclusion type insect viruses 32, 135
Smithiavirus 134
sodium-potassium ratio in hemolymph 110
soluble antigen measurement as assay for plant virus multiplication in insects 95
species specificity of nuclear and cytoplasmic polyhedrosis viruses 114
specificity of arthropod viruses 151
spectra of virulence of non-inclusion type viruses of invertebrates 25, 26, 27, 28
spherical virions 1

St. Louis encephalitis virus 39, 40, 43, 45, 46, 49, 51, 52, 54
— — — —, infection of birds 49
stabilized state of sigma virus infection 85, 86, 87
Steinhausia virus 134
stomach cells, changes in nuclei in virus infected aphids 102, 105
— —, enlargement of nuclei in vectors of leafroll and pea enation mosaic virus 102, 105
Streptococcus apis 26
structure of arthropod viruses 130
— inclusion-type insect viruses 1, 135
— non-inclusion-type insect viruses 32, 135
subgroups of arbovirus group B 39
suckling mice, use for isolation of arboviruses 50
sunlight, influence on viral instability 149
super-activation of latent virus infections 153
survival of inclusion-type insect viruses 13
— — — — in the hosts without pathologic or cytopathic changes 12
— nuclear-polyhedrosis viruses 12
susceptibility in *Lepidoptera* to nuclear-polyhedrosis viruses of different species 15
— of hemolymph cells, importance for high biological specificity of leafhopper vectors 102
— larvae to insect viruses of inclusion-type, variation of 14
— to nuclear-polyhedrosis virus, increase by blocking the host defense mechanism 18
— — —, — — the phagocytes with carbon particles 18
— — —, — heat-inactivated granulosis virus 18
— polyhedrosis viruses in different larval stages 14, 15
symptomatology of arbovirus infection in vertebrates 49
synopsis of viral effectiveness, safety and feasibility 158
systemic infection of an aphid by a plant pathogenic virus 104
— invasion of an insect vector by a plant-pathogenic virus 96
— — leafhopper by wound tumor virus 98
Tacaribe group of viruses 43, 49, 50
temperature, importance in development of epizootics 20
—, inducing development of resistance against viruses 151
—, influence on CO_2-sensitivity 72
—, — sigma virus multiplication 72
—, — — — transmission 81, 86
—, — transmission of rho-infection 83
—, — — ultra-rho infection 83, 86
—, — viral activity 149
— of incubation and rate of virus development 45
— resistant (Tr) sigma virus strains 70, 73
— sensitive (Ts) sigma virus strains 70, 72
teratogenicity of viral insecticides 153
TE virus, primary tick tissue cultures as sensitive assay system for low concentrations 124
— — strain Hypr, comparison of sensitivity of primary tick tissue cultures and chick embryo cell cultures 124
— — — —, multiplication in tissue cultures of *Dermacentor marginatus* and *Hyalomma asiaticum* 124
Thaumetopoea pityocampa 147
Thaumetopoeidae 17, 147
thermosensitivity of sigma virus infection at each stage of oogenesis 82
thin sections of vector cells, examination for detection of plant virus multiplication in insects 95, 98, 102
thoracic tissue culture of the variegated cutworm *Peridroma saucia* 109
thymidine, effect on sigma virus multiplication 65
tick-borne virus complex of arbovirus group B 41, 42
tick tissue cultures, studies on arboviruses in 123
ticks as vectors of arboviruses 42
timing of application of living insecticides 145
Tineidae 17
Tipula iridescent non-inclusion virus 142, 154, 156, 158
— — — —, experimental transmission 142

Tipula iridescent non-inclusion virus, safety to various organisms 154
— *paludosa*, larvae of 27, 33
tissue affinity of non-inclusion type viruses of invertebrates 30, 116, 134
tissue-cell cultures as source of insect viruses 158
— cultures, use for isolation of arboviruses 50, 51
— specificity of insect viruses in vitro 116
— — — — in vivo 116
tissues, susceptible to the inclusion-type insect viruses 8, 9, 116, 134
titers of sigma virus during different stages of flies development 66
— — — — — of infection 66
titration of sigma virus 64
tobacco mosaic virus 133
tolerances, prescribing the permitted amount of residue levels on crops after insecticide treatment 153
Tortricidae 4, 17
toxicity of viral insecticides 153
tracheae 98, 105, 116
tracheoblasts 98
transmission experiments with the virus of the flaccidity of silkworms 26
— of arboviruses 38, 43
— sigma virus 76
— — —, influence of fly and virus mutations 76
— trials with a virus in European foulbrood of the bee 26
— — the virus of lethargy of *Melolontha melolontha* 28
— — — paralysis of crabs 26
— — — — of crickets 25
— — — transparency of *Scarabeidae* 25
— — virosis of *Acarina* 27
transovarial infection of rice leaf-hopper by rice dwarf virus 95
— passage as proof of multiplication of plant-pathogenic viruses in insect vectors 95
— — viruses in arthropods 45, 46, 59, 60, 77, 95, 121
— — — insects, usefullness of tissue cultures for demonstration of 122
transovum transmission of virus 152
transparency of *Scarabeidae* 25
— —, pathological findings 25
transparent disease of *Melolontha melolontha* 33
transplacental transmission of arboviruses in vertebrate hosts 47, 48
transtadial virus transmission 152
trehalose, presence in hemolymph 110
Trichoplusia ni 15, 144, 151, 152
— —, susceptibility to nuclear polyhedrosis virus 15, 144
Trivittatus virus 39
trypsin, damage of insect cells by 109
tubular structures surrounding plant pathogenic viruses in infected insect vectors 102
tussock moth, tissue cultures from 111
types of arthropod viruses 1, 24, 59, 129, 134
— — —, propagation of 156
types of insect viruses, causing inclusion bodies 1, 134, 135

ultra-rho behavior of sigma virus infection 61, 69, 84
— — infection 71
— — —, effect of temperature on the transmission of 83, 86
— — —, influence of temperature on 75
— — —, transmission of infection without any extractable infectious material 86
ungrouped arboviruses 39
unusual aspects of sigma virus infection 83, 84
U.S. bat salivary gland viruses 39
U.S. Department of Interior reviews on the registration-application for its role in human health and wildlife conservation 153
useful invertebrates, hygiene of 34
UV light, influence on viral instability 149

vaccination against arbovirus infection 53
VAGO medium 110
Vagoia virus 134, 139
valence, i.e. fraction of sigma virus-bearing sperm 76
Vanesa cardui cytoplasmic polyhedrosis virus transmitted to *Hymenoptera* and *Lepidoptera* 151

variability in response to granulosis virus and genetic constitution of different populations 16
— — nuclear-polyhedrosis virus of *Heliothis* species 16
— of host response to infection with inclusion-type insect viruses 15, 16
vector reduction used for the control of adenoviruses 53
vectors for plant viruses 94
Venezuelan equine encephalitis virus 52
— — — — in wild rodent hosts 49
Vero cell culture used for isolation of arboviruses 51
vertebrate hosts of arboviruses 46, 48, 49, 50
vertebrate safety test of viral insecticides 153
vertical transmission of inclusion-type insect viruses 8, 10
vesicular stomatitis virus 43, 91
— — —, multiplication in *Drosophila melanogaster* 91
— — —, transmission by *Aedes aegypti* 43
viability of inclusion-type insect viruses outside their hosts 13
viral disease of *Cirphis unipuncta* 25
— insecticides, production and industrial feasibility 155
viremia after infection with arboviruses in different vertebrate hosts 47
— in hibernating animals 47
— snakes 47, 48
VIREX 157
virions containing DNA nucleo-protein 1, 27, 135
— — RNA nucleoptotein 2, 25, 135
— of the nuclear-polyhedrosis and granulosis viruses, similarity 6, 135
— wound tumor virus, formation 102
VIRON/H^{tm} 157
viroplasmic matrix, sites of virus multiplication 98
viroplasms 95, 99, 100, 101, 102
— in plant and insect host cells, assembly of wound tumor virus particles in 101
— wound tumor virus infected plant cells 100
— of wound tumor virus, similarity to those of other viruses of 102
viroplasms, similarity to virogenetic network 95
virosis of *Acarina* 27
— —, pathological findings 27
— —, spectrum of virulence 27
virulence of cytoplasmic polyhedrosis viruses 16
— nuclear-polyhedrosis viruses 18
— — — in different species of the genus *Malacosoma* 15
virus-bacterial interactions in European foulbrood of the bee 27
— behavior in the arthropod vector 43, 44
— -host-relationship in inclusion-type insect viruses 13
— infections of insects 1, 24, 38, 59, 94, 129, 134
— interactions in species infected with granulosis and nuclear-polyhedrosis viruses 17
— subcommittee on bacteriological nomenclature, insect virus study group of 142
— titers in infection of nuclear polyhedrosis virus 8, 113
— units per oogonial cell in sigma and rho infections 83
viruses associated with insects 1, 24, 38, 59, 94, 108, 129, 134
—, available for field tests 159
—, distribution among arthropods 140
— grown in living insects or tissue-cell cultures 111, 158
— isolated from arthropod pests 141
—, living insecticides 129
—, pathogenic to insects, in vitro studies 111, 158

West Nile fever virus 39, 123
western encephalitis virus 49, 52
— equine encephalomyelitis virus, survival and multiplication in *Aedes aegypti* tissue cultures 123
— X-disease virus, cytopathogenic and systemic changes in its vector, caused by 96
— — —, injurious effect on its vector, *Colladonus montanus* 96
— — — of stone fruits 96
wild vertebrate host reduction used for the control of arboviruses 54

wildlife conservation, role of the U.S. Department of Interior review on the registration-application for 153
wound tumor virions in cytoplasm of ganglion cells and in glial cells 98
— — viroplasms, similarity to those of other viruses 102
— — virus, accumulation in fat body tissue 96
— — —, assembly sites in the insect vector 98
— — —, concentration in tissue cultures 122
— — —, detection in tissue culture cells by immunofluorescent staining 122
— — —, development in fat body tissue of insect vectors 102
— — —, — insect vectors after abdominal inoculation 102
— — —, — primary tissue cultures of the leafhopper *Agallia constricta* 122
— — —, distribution in central nervous system of *Agallia constricta* 98
wound tumor virus, distribution in tissue of leafhopper vector 97
— — —, electron micrographs 97, 99, 100, 101
— — —, multiplication in the leafhopper vector *Agallia constricta* 95
— — —, — tissue cultures demonstrated by a series of passages 122
— — —, sequential stages of development in insect vectors 102
— — —, systemic invasion of the mechanically infected insect host 102
WTV 95, 97, 99, 100, 101, 122

yeast extract as supplement in medium for insect tissue cultures 110
yellow fever vaccine, 17 D strain 53
— — virus 39, 52

Zeiraphera diniana 14
— — response to granulosis virus of geographically separated populations 16
zoophagous insects, medium requirements for tissue cultures of 110

Universitätsdruckerei H. Stürtz AG Würzburg

MIX
Papier aus verantwortungsvollen Quellen
Paper from responsible sources
FSC® C105338

If you have any concerns about our products,
you can contact us on
ProductSafety@springernature.com

In case Publisher is established outside the EU,
the EU authorized representative is:
Springer Nature Customer Service Center GmbH
Europaplatz 3, 69115 Heidelberg, Germany

Printed by Libri Plureos GmbH
in Hamburg, Germany